# Testimonios
## Stories of
## Latinx and Hispanic
## Mathematicians

# Testimonios
## Stories of Latinx and Hispanic Mathematicians

Edited by

Pamela E. Harris

Alicia Prieto-Langarica

Vanessa Rivera Quiñones

Luis Sordo Vieira

Rosaura Uscanga

Andrés R. Vindas Meléndez

Illustrated by Ana Valle

*Produced and Distributed by*
*The American Mathematical Society and The Mathematical Association of America*

2020 *Mathematics Subject Classification.* Primary 00-XX, 01-XX, 01A70, 97-XX.

Illustrations of the authors and editors created by Ana Valle.

For additional information and updates on this book, visit
**www.ams.org/bookpages/clrm-67**

**Library of Congress Cataloging-in-Publication Data**

Names: Harris, Pamela E., 1983- editor. | Prieto Langarica, Alicia, 1983- editor. | Rivera
  Quiñones, Vanessa, 1990- editor. | Sordo Vieira, Luis Alfonso, 1991- editor. | Uscanga Lomeli,
  Rosaura, 1989- editor. | Vindas Meléndez, Andrés Rodolfo, 1992- editor. | Valle, Ana, 1995-
  illustrator.
Title: Testimonios : stories of Latinx and Hispanic mathematicians / Pamela E. Harris, Alicia
  Prieto-Langarica, Vanessa Rivera Quiñones, Luis Sordo Vieira, Rosaura Uscanga, Andrés R.
  Vindas Meléndez, editors ; illustrated by Ana Valle.
Description: Providence, Rhode Island : American Mathematical Society/Mathematical Associa-
  tion of America, [2021] | Series: Classroom resource materials, 1557-5918 ; volume 67 |
Identifiers: LCCN 2021016982 | ISBN 9781470466572 (paperback) | 9781470467159 (ebook)
Subjects: LCSH: Hispanic American mathematicians–Biography. | Mathematics–Study and
  teaching–Social aspects–United States. | AMS: General and overarching topics; collections.
  | History and biography. | History and biography – History of mathematics and mathemati-
  cians – Biographies, obituaries, personalia, bibliographies. | Mathematics education.
Classification: LCC QA28 .T47 2021 | DDC 510.92/2–dc23
LC record available at https://lccn.loc.gov/2021016982

*For those who came before and for those who follow.*
*A window to see our stories. A mirror to see ourselves.*

# Contents

Dr. James A. Mendoza Álvarez's *testimonio* begins with the story of his mother, Olga Mendoza, whose life takes us through a time in the twentieth century when signs stating "Mexicans not allowed" were commonplace in Texas. Olga's struggles, wisdom, and influence, not only illustrate the plight of people of Mexican descent, but have also been a major driving force in Dr. Álvarez's life and in his professional trajectory. After his *testimonio*, Dr. Álvarez gives advice for those of us within academia whose work aims to address the underrepresentation of Latinxs and Hispanics in the mathematical sciences.

Dr. Federico Ardila Mantilla's *testimonio* begins with the story of his family to show us where he comes from. He guides us through his mathematical journey, beginning with learning to love math as a result of participating in the math olympiads, then learning the power of working in collaboration and community, and finding his mathematical home at SFSU. Dr. Ardila Mantilla shares some of his difficult experiences, teaches us about life balance, and encourages us to not close doors on ourselves, and instead open them and leave them open for those who come after us.

Dr. Selenne Bañuelos' *testimonio* is about family, community, culture, and their importance in cultivating talent. As a child of immigrants, she learned from her parents the value of hard work and the power of education. She always believed she could, not in spite of being a Chicana woman, but because of it. Her story is one of overcoming adversity with the strength of discipline, hard work, and a very clear goal.

## Dr. Erika Tatiana Camacho                                         **41**

Dr. Erika Tatiana Camacho's *testimonio* is centered around the importance of paying it forward, mentorship, and changing the world through mathematics. Dr. Camacho tells us how her mentor-mentee relationship with Jaime Escalante changed her life, highlighting the importance of such relationships. She shares some of her experiences with overt discrimination. Dr. Camacho then tells us about her amazing work developing mechanistic models to aid in understanding causes of blindness, and shares with us her experiences in driving the Latinx community forward through her efforts.

## Dr. Anastasia Chavez                                              **51**

Dr. Anastasia Chavez' *testimonio* brings us a story of perseverance in the face of discrimination. She tells us about how support is an essential need to overcome systemic barriers that exist in academia. Her story is an inspiring one—riddled with difficulties from raising a family in the middle of graduate school and a lack of support from gatekeepers—but ending with becoming an outstanding researcher in algebraic combinatorics who does not forget to help and inspire those who need it.

## Dr. Minerva Cordero                                               **59**

Dr. Minerva Cordero's *testimonio* walks us through her early passion for mathematics, her family life, and the mentors who inspired her to pursue a career in mathematics. Her resilience through family tragedy became her motivation to continue to pursue her passion for mathematics. Dr. Cordero believes in the value of both mathematics and diversity in advancing society. Through her teaching, mentorship, and research, she wants to encourage the interest of underrepresented minorities and women in Science, Technology, Engineering, and Mathematics (STEM).

## Dr. Ricardo Cortez                                                **71**

Dr. Ricardo Cortez's *testimonio* is a story about family, perseverance through hardship, and giving back to the community through service. After being raised in El Salvador, he pursued bachelor's degrees in engineering and mathematics in the United States. It was through the advice of mentors and peers that he decided to pursue a PhD. During his career, Dr. Cortez has been involved in many efforts to address the issues faced by underrepresented students in mathematics. Lastly, he emphasizes the importance of maintaining your own professional advancement to ensure that these efforts are effective.

## Dr. Hortensia Soto                                                            191

Dr. Hortensia Soto's *testimonio* is shaped by hard work. Rising before sunrise as a child through her youth to help her family farm, she developed a strong work ethic, which has served her well in every aspect of her life. This, together with her resilient spirit, helped convince her mother to allow her to leave home and attend college. Throughout her story, Dr. Soto highlights those on her "gratitude list" which is comprised of key individuals who have helped her and her family. Her gratitude list reminds us of the value of community. After her *testimonio*, Dr. Soto gives brief advice to both students and mentors.

## Dr. Roberto Soto                                                             203

Dr. Roberto Soto's *testimonio* takes us through an "unexpected journey" through the world of numbers that almost did not happen. How such travels and experiences in experimental mathematics collide with mentors throughout life is a truly incredible story of how diverse all of our paths are on the way to becoming professional mathematicians.

## Dr. Richard A. Tapia                                                          213

Dr. Richard A. Tapia's *testimonio* is a unique story of resilience and achievement. Dr. Tapia rose from humble beginnings to attain the greatest heights of an academic life, or—as he puts it—to run with the big dogs. Dr. Tapia touches on significant tragedies in his life story and reminds us throughout how critical it is to have a strong support system of family and friends. He shares his work mentoring underrepresented minority students and his views regarding the importance of enlarging the pool of mentors with similar backgrounds. He reminds us that improving the future of underrepresented minorities begins by considering them not as a unidimensional group, but rather as a multifaceted set of talented individuals. Dr. Tapia touches on his many outreach efforts, including grant-funded efforts and the co-founding of SACNAS. His *testimonio* takes us from Los Angeles to The White House, where he received the National Medal of Science from President Barack Obama.

## Dr. Tatiana Toro                                          225

In Dr. Tatiana Toro's *testimonio* we learn of the challenges faced by her family given the difficult political climate in Colombia in the 1950s. She shares her experiences while pursuing a mathematics career during a time when her moving to study in France, and later to the United States, was thought to lead a young woman to "the road to perdition." Her perseverance and determination helped her push through to complete her studies and become an award-winning mathematician. Among her numerous contributions, she shares with us her work in the organization of the Latinxs in the Mathematical Sciences Conference, the first of its kind in the United States.

## Dr. Anthony Várilly-Alvarado                            235

Dr. Anthony Várilly-Alvarado's *testimonio*, walks us through his family history from Brazil to Costa Rica, where his own story began. Growing up he had a passion for the abstract and it translated into pursuing mathematics as his career. His research interests lie in the area of arithmetic geometry, where he studies diophantine equations and currently applies his expertise in collaboration with earth scientists.

## Dr. Mariel Vázquez Melken                               249

Dr. Mariel Vázquez Melken's *testimonio* is a story of love for her city, her roots, and for mathematics. Dr. Melken details growing up in the largest city in the world, immersed in its incredibly rich culture and the struggle of its overpopulation. As a teenager she fell in love with mathematics and microbiology. She chose math thinking that she had to choose one of the two. Fortunately in college she realized she could study both. She is now one of the leading mathematical biologists in the world. Her *testimonio* also talks about being a woman in a male-dominated field and, more importantly, how she has learned to survive and fight for others.

## Dr. William Yslas Vélez                                 259

Dr. William "Bill" Vélez's *testimonio* is a story centered around resilience, family, and his partnership with his wife Bernice. Despite growing up in poverty, Dr. Vélez shows us how the support of his family was an important component for his success. He shares the role of Mexican-American culture in his journey and how his mother was a key figure in his education. He also discusses his experience as the first Chicano hired in a tenure-track position at the mathematics department at the University of Arizona, and how he found his voice in the mathematical community. Dr. Vélez continues his efforts to support his students and increase minority representation in mathematics through active participation in professional organizations such as SACNAS and the Math Alliance.

Dr. María Cristina Villalobos' *testimonio* frames her story through an important life lesson of taking the initiative; a lesson she now passes on to her students. She recounts the ways in which "taking the initiative" helped her find mentors, pursue new opportunities, and how it led to becoming a driving force in broadening the STEM participation of women and minorities. Being the recipient of a 2019 Presidential Award for Excellence in Science, Mathematics, and Engineering Mentoring, a top honor awarded by the White House, Dr. María Cristina Villalobos is proof that it takes initiative to reach our goals.

Illustrations of the authors created by Ana Valle.

# Lathisms
## Latinxs and Hispanics in the Mathematical Sciences

It is often the case that mathematical proofs and theorems seem to magically appear out of thin air—as if in that magical moment of epiphany, the proof wrote itself in its beautiful form. Of course, for most of us, this is only an illusion—theorems occur after mental marathons, struggles, and refinement. Books also appear to us as if they fall from the heavens, often lacking a story of how the book came to fruition. We decided that it would be a disservice to not tell the story of how this book came to fruition, for every mathematical story has people behind it. Thus, we decided to begin this book of *Testimonios* with the story of Lathisms: Latinxs and Hispanics in the Mathematical Sciences. It is the editors' personal *testimonios* of a mathematical family brought together by a deep need to be seen and accepted as our authentic selves within mathematical spaces. We consider this book to be a tangible source of inspiration, not only for those who lack a mathematical family and those rising through the ranks, but for ourselves.

## A Chance Meeting – Pamela and Alicia

**Pamela Harris**. I completed my PhD studies in 2012 at the University of Wisconsin Milwaukee, and I later learned that Alicia completed her PhD studies the same year at University of Texas at Arlington. I worked on problems in combinatorial representation theory and Alicia on problems in mathematical biology. It is truly an example of mathematical and personal fate that we would meet each other, as we couldn't be further apart in our mathematical interests. Yet, this is the beauty of being MAA Project NExT (New Experiences in Teaching) fellows. I was a Silver '12 and Alicia a Brown '13 and we both attended MathFest 2013 in Hartford, Connecticut. Although the story of our meeting is something I prefer to tell over mezcal and tacos, I am happy to share it here because of the deep impact it had on us and in the many friendships we have cultivated since our meeting.

During MathFest 2013, I co-organized a panel on how to write letters of recommendations; something that no one ever taught me to do as a graduate student, but that I needed to learn as I was now a professor. Setting-up for the panel, I realized that none of us as organizers had remembered to bring a clicker to advance the panelists' slides. I left the room in a hurry looking for Dr. Aparna Higgins, who at the time was the Director of Project NExT. I was sure that Aparna would be able to lend me a clicker. When I found Aparna by the registration table, she thankfully had a clicker. I fondly remember that as we fidgeted with the batteries ensuring they were placed in the correct orientation, Aparna took advantage of the opportunity to teach me that one should keep the batteries out of the clicker when not in use so that the batteries are not dead on arrival. With the working clicker ready, I quickly walked away to return to my panel, without a clue that the next few minutes of my life would lead to finding my mathematical soulmate.

**Alicia Prieto**. As a recent immigrant from Mexico, I had always had a hard time getting used to being "the only one" at every math conference or meeting. After a while, I made it a point to find any person of color that I could bond with. After realizing that among the 80+ Brown '13 fellows there were exactly eight nonwhite participants, only four Latinxs of which only one was U.S. born, I decided to confront Aparna about this lack of representation within the Project NExT Fellows. This was the question I posed Aparna, who just minutes prior had handed Harris[1] a clicker, and she replied: "No, there are more! There is a Pamela Harris!" pointing to Harris who was about 30 feet away. With that last name, I was skeptical that Harris could be Latina. So when Aparna pointed her out in the crowd, as she was walking back toward her panel, and after I realized that Harris did "look" Mexican, I softly called out (in Mexican full-on volume) toward Harris's direction: "Brown people!! Brown peopleeeeeeeeeeee!" As Harris turned to look toward me to see who was screaming "Brown people," and realizing that in fact I was shouting at her, she turned and saw me sprinting toward her. Harris stopped and waited for me to get closer and that's when I began my inquisition:

> Alicia: *Where are you from?*
> Pamela: *México*
> Alicia: *¿Hablas Español?*
> Pamela: Si.
> Alicia: *Y ¿de donde en Mexico?*
> Pamela: *Guadalajara.*
> Alicia: *Yo también!*
> Pamela: *Ok. Sorry, but I have to go because my panel is about to start. Maybe we can go to lunch afterwards?*

I quickly scribbled my phone number on a scrap of paper to share with Harris; she still has that scrap paper and it is pictured on the next page. As Harris went off to her panel, I reflected on how happy I was to find someone who not only looked like me, but who was born in the same Mexican city of Guadalajara, Jalisco. By the end of our lunch that day we discovered that we were born just three days apart in the same city, we are the oldest of three siblings (one sister and one brother), and both of our mothers have the same name: Rosa. We also quickly discovered that we had another very important thing in common: a terrible thirst for a family within the mathematical community|a family of people who shared our culture, values, and beliefs. A family we believed existed, but which often felt so invisible. That chance encounter was the beginning of the Lathisms family.

---

[1]  This is the official nickname for Pamela.

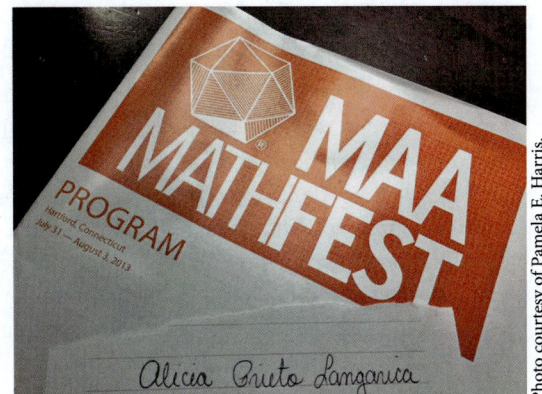

Scrap of paper where Alicia gave Pamela her phone number along with a copy of the program for MAA MathFest 2013.

Photo courtesy of Pamela E. Harris.

## The Founding of Lathisms

**Pamela Harris**. Many good ideas have begun from conversations and discussions via social media. Lathisms is one such idea which began in the spring of 2016 when on social media someone asked if there was a repository or a list of minority mathematicians. Someone pointed to the SACNAS biography project, yet this only had a few mathematicians. Motivated by this, in private conversations with Alicia, Alexander Díaz-López, and Gabriel Sosa, we decided that if no such website existed, we would create it. We were especially excited for the timing since it would allow us to unveil this website during Hispanic Heritage Month that upcoming fall.

This was the birth of Lathisms: Latinxs and Hispanics in the Mathematical Sciences in 2016. We began Lathisms with the idea of bringing visibility to the vibrant and active Latinx and Hispanic community within the mathematical sciences by creating a Hispanic Heritage Month calendar (running yearly during National Hispanic Heritage month, from September 15–October 15) in which a featured mathematician was uncovered daily. Our primary goal was to provide an accessible platform that featured the multifaceted and diverse nature of the Latinx and Hispanic mathematical community and which would inspire younger generations of mathematicians. This work expanded to collaborations with the American Mathematical Society for a poster, and with the financial support of the Mathematical Association of America we were able to expand our efforts to include podcast interviews hosted by Evelyn Lamb.

**Alicia Prieto**. The work of Lathisms in those initial years came with a lot of feedback from the community at large. Most of this feedback was quite positive, but some came with a warning to us about the time commitment it takes to make such initiatives last. This was well-intended seeing how we were all early-career mathematicians, and those sharing their concern were right to think that we might fall behind on our research and possibly live through some negative repercussions later on in our careers.

Harris and I are the first to admit that the workload in running Lathisms was intense, especially as there were only four of us involved during those initial years. The support from the MAA in the form of a Tensor SUMMA grant alleviated some of the workload as we were able to hire a web developer, Richard Diaz, but as Alexander stepped down to move onto other projects and later Gabriel; Harris and I needed to reassess the workload and bring in additional people with a passion for advancing the Lathisms mission and vision and who would help us lead the organization. Hence, in 2018 we created a Lathisms Leadership Team and a Lathisms Junior Associate Team.

**Pamela Harris**. With the need for additional help to run Lathisms, Alicia and I discussed who to include in the Lathisms Leadership Team. We both agreed that we needed someone who was dependable and passionate about addressing issues of underrepresentation of Latinxs and Hispanics in the mathematical sciences. Naturally, Luis Sordo Vieira came to mind given his move to research in medicine following his completion of a PhD in number theory. After working together for a few months, we realized we needed more help. This sparked the idea of putting together a Junior Associates Team, made up of rising stars who shared our values and who could help shape the future of Lathisms. Our first goal was filling a need for someone who had a background in scientific writing and a network within applied mathematics, and luckily we knew Vanessa Rivera Quiñones who came highly recommended by Alexander Diaz-Lopez and had established a track record in her work.

**Alicia Prieto**. New to the team was another rising star within our community; Andrés Vindas Meléndez, whose trajectory Pamela knew of from years prior, brought energy and renewed passion for highlighting the broad diversity within the Lathisms community. Finally, we knew we needed an expert in mathematics education, especially as the 2019 Lathisms calendar would focus on math educators. Fortunately, I knew one of the BEST mathematics education researchers, who happened to be another amazing Latina, and who happened to be one of my best friends: Rosaura Uscanga Lomelí. The addition of Vanessa, Andrés and Rosaura to the Lathisms Team has made the team awesome.

## The Lathisms Team

**Luis Sordo Vieira**. I graduated from the University of Kentucky in 2017, completing my undergraduate studies at Wayne State University. As a Venezuelan-American mathematician, I, like Alicia, also experienced being "the only one" in several spaces, as this has often felt like the rule rather than the exception. Although I had met Pamela at the Latinos in the Mathematical Sciences Conference at the Institute for Pure and Applied Mathematics in 2015, Lathisms was where our friendship really started. I found a mathematical older sister, not in the sense of an academic co-descendant, but rather in the sense of an additional familial older sister, a person I could trust who would tell me when I messed up, cherish my successes, and give me a hand when I struggled. When I decided to change focus from my PhD work in number theory to applying mathematics to medicine in my late years of graduate school, I remember my doubts and fears of failing fading away as I told Pamela about my plans. Pamela displayed an honest joy—a joy that many reserve for their own success.

A year after graduation, Alicia (who by then I had already met and admired her quick wits and fearless will to speak out) and Pamela invited me to join the Lathisms family as part of the Leadership Team. I accepted without hesitation, as it is a rare opportunity to work side-to-side with your family on something that you believe to have a deep impact on your community. And the Lathisms family kept expanding. I find a great sense of joy in contacting the Lathisms calendar nominees, reading why they were nominated, and in reading and editing their stories. Reading their stories and working with the rest of the Lathisms team to highlight all of these outstanding accomplishments on either the website, articles, or this book, served not only as a source of inspiration for the pursing of new mathematics—but also as a source of motivation to keep striving for a more equitable society.

**Vanessa Rivera Quiñones**. Born and raised in Puerto Rico, I was always excited to meet other Latinx/Hispanic mathematicians. As an undergraduate, I studied at the University of Puerto Rico at Río Piedras and met Alexander Díaz-Lopez during a Calculus competition. I saw in Alexander a contemporary role model and was very excited about his work with Lathisms. However, joining Lathisms came a while later, when Pamela invited me to join the team. At the time, I was a graduate student at the University of Illinois at Urbana-Champaign. I had met Luis and Andrés at the Latinxs in the Mathematical Conference in 2018, and seen the powerful work done by the co-founders. So, I jumped at the opportunity to collaborate with the team.

It has been all I ever hoped for and more. I am inspired by each of the members, and how they bring their full selves to this life-transforming work. I've learned so much from Pamela, Alicia, Luis, Rosaura, and Andrés. It's through their passion, dedication, and hard-work that Lathisms has been able to grow and become a family. By featuring and sharing the personal and professional stories of Latinx/Hispanic mathematicians through so many avenues (i.e., our website, posters, podcasts, and articles), we've painted a rich picture of our honorees and our community. For me, working with the Lathisms team has been a way to honor those that are paving the way to make mathematics a home for us.

**Andrés Vindas Meléndez**. I first met Pamela during the 2012 SACNAS National Conference, where Pamela was a poster judge and I was presenting my undergraduate research. We corresponded by email after the conference with intentions to work on mathematics together, but life happened and mathematical collaborations did not come until years later. After completing my undergraduate studies at UC Berkeley, I pursued a master's degree at San Francisco State University, where I found a community that shared my passions for both mathematics and social justice. At San Francisco State, I co-founded and co-organized the Distinguished Women in Mathematics Lecture Series in the Fall of 2016. Fortunately, Pamela was attending an American Institute of Mathematics workshop in nearby San José and accepted our invitation to be our first distinguished speaker. Since then we have attended and bonded at several national and international conferences, collaborated on research and organizing conference sessions. I met Luis at the 2015 Latinos in Math conference. It was encouraging to meet other Latinx graduate students at the conference and this is where Luis shared with me his experiences as a graduate student at the University of Kentucky; I would go on to attend the University of Kentucky for my PhD, exemplifying the power of connections. I had the pleasure of meeting Alicia (after having heard amazing things about her) at the 2018 Critical Issues in Mathematics Education Conference hosted by the Mathematical Sciences Research Institute at Berkeley. It felt as if I had known Alicia forever. As mentioned by Vanessa above, we met in person at the second iteration of the Latinos in Math conference in 2018. As a fun fact, I had actually emailed Vanessa for help on a graduate fellowship a few years before after noticing her name on the roster of award recipients. While I have not had the pleasure of meeting Rosaura in person, I value her presence on the team and admire her contributions and knowledge of mathematics education.

When I was asked to join the Lathisms team, as a Junior Associate, I accepted immediately since I knew the team is amazing. In a short time we have connected and have created very close bonds. Our different personalities, qualities, and abilities, makes the team work so well! Since joining the Lathisms team, I have helped expand the work and network of Latinx mathematicians. My goals within Lathisms is to facilitate the professional development of Latinx mathematicians and their supporters to create a space where everyone feels comfortable and supported so that they can thrive in whatever mathematical pursuits and directions they take.

**Rosaura Uscanga Lomelí**. Alicia and I met in 2008 back when I was in my first year of my undergraduate studies and she was working on her PhD at the University of Texas at Arlington. There were not many Hispanics in the mathematics department so I noticed Alicia right away, but as a first year student in college, I was scared to approach her. If you know Alicia, you know that she makes a significant effort to build a network with fellow Black and Brown mathematicians, so she noticed me one day at math club and made sure to come introduce herself (she asked me who I was, where I was from, and more importantly for her, if I spoke Spanish). We found out we were both from México which gave us a shared perspective on life in the U.S., and being around her made me feel close to home. Right away I felt a connection with her! She has been a constant source of inspiration and support in my life. For the rest of my undergraduate education, we met weekly for lunch and for walks around campus. She quickly became an important part of my life and I was excited for her to get to know my family and for them to get to know her. So I invited her over to my house (I lived with my parents at the time) to meet my whole family and they loved her as much as I do. From then on, she knew she was always welcome at my house and at any family gatherings; she was part of the family. She became like my older sister (although she says younger, because she claims I am more mature but I disagree).

In light of our deep connection, I happily accepted the invitation to join Lathisms in 2019! I was honored that they thought of me to help out with their mission. At the time, I only knew of Pamela because I had heard a lot about her from Alicia and her story had inspired me—the stories I had heard about her life resonated with me. So I thought there was nothing better than getting the opportunity to work with two amazing Latina mathematicians. Once I became part of the team, I "met" (online) Vanessa, Andrés, and Luis. While I have not yet had the pleasure of meeting them in person, these amazing individuals have made a huge impact in my life and inspire me each and every day. They are passionate about everything they do and are dedicated to making Lathisms better each year and to make the road easier for Latinx and Hispanic students who aspire to become mathematicians. I had never before been surrounded by so many Hispanics in mathematics and I've got to admit that it is pretty awesome. Working at Lathisms does not feel like work, our weekly meetings feel just like spending time with family.

# Testimonios: Stories of Latinx and Hispanic Mathematicians

Mathematics is not created nor discovered in a vacuum. Inherent to mathematical progress is the stories of the people behind it. So as we reached the fifth anniversary of Lathisms, we wanted to highlight the power of stories about our community and family. We compiled this book because we thought that seeing the *testimonios* of inspiring mathematicians could help further the growth and the brilliance of the community of Latinx and Hispanic mathematicians. Our goal has always been to inspire younger generations of Latinx and Hispanic mathematicians, so that they may see themselves reflected in these stories, and so they may learn that we stand on the shoulders of giants; inspiring the next Tapias and Toros of mathematics. This book also highlights rising-stars in our community. Their lives and journeys will inspire current undergraduate and graduate students who often seek a role model who shares their language, culture, and heritage, and who are clearing the path in front of them so they can reach further than ever before. Moreover, these stories talk about the American experience at large. We seek to inspire anyone who has or wants to have a career in mathematics or any other STEM field—particularly those from underrepresented groups. In general, we want to reach a wide variety of individuals at all different stages in their careers. We hope this book helps people outside of our community, specially those who want to be allies or mentors, realize the particular challenges faced by minoritized populations and we offer, in the many examples, a window to potential barriers and to ways in which everyone can help overcome them.

In addition, this book was also for ourselves. As we read every *testimonio*, we relived those stories and imagined the impact these narratives would have had on our younger selves and on our trajectories in the mathematical community. How less isolated, lonely, and lost would we have felt as some of us immigrated to a new country, learned a new language, or failed graduate school examinations. Knowing we are not the only ones to have struggled and who have overcome would have helped us tremendously in those times of need. It is because of this that we consider this book a true success, for the impact it has had on us is immeasurable. These stories have truly inspired us. The great efforts to overcome challenges, both personal and systemic, and the resilience and devotion of outstanding researchers and educators showcased through these stories have energized us in an extremely difficult year.[2]

We end by extending our thanks to Ana Valle for the beautiful illustrations accompanying each chapter, and by expressing our deep gratitude to all of those who contributed to this book. We thank you for sharing your stories, your lives, your beings, fully and authentically. We want to acknowledge that, for many, this was a cathartic, and sometimes, emotionally painful and time consuming process. We are confident your *testimonios* will inspire all within the mathematical community to pay attention to the struggles, inequalities, and problems that we, as a community and as individuals, should strive to resolve in order to improve the experience of those marginalized in mathematics, including Latinxs and Hispanics. Our future is brighter because of your words and your work.

Con mucho amor y respeto,

Pamela E. Harris
Alicia Prieto Langarica
Vanessa Rivera Quiñones
Luis Sordo Vieira
Rosaura Uscanga Lomelí
Andrés R. Vindas Meléndez

---

[2] This book was written and edited during the global COVID-19 pandemic.

The Lathisms Team.

## Agradecimientos y Dedicatorias / Acknowledgements and Appreciation

**Pamela**. *Toda historia tiene un comienzo, y para mi esta empieza con mis gracias a mis padres, Rosa y Jorge, por todos los sacrificios que han hecho para que yo tuviera la oportunidad de poder realizar mis sueños. Toda mi vida seguiré tratando de hacerlos orgullosos por que quiero que sepan que sus sacrificios no fueron en vano. Para mi hermana y mi hermano, Ana y Jorge, les agradesco toda una viva de inspiración. Para mi hija Akira, mi sueño para ti es que encuentres tu pasión y que continues trabajando para crear un mundo mas justo. Para mi pareja, Jamual, mi amor desde los quince años. No tengo palabras para agradecerte todo lo que siempre has hecho por mi y por nuestra familia. Lo bueno que he podido contribuir a este mundo es por que siempre me as apoyado y querido. Te amo!*

Every story has a beginning, and for me it begins with my thanks to my parents, Rosa and Jorge, for all of the sacrifices they have made so that I had the opportunity to make my dreams come true. Throughout my life I will continue trying to make you proud because I want you to know that your sacrifices were not in vain. For my sister and my brother, Ana and Jorge, I thank you for a lifetime of inspiration. For my daughter Akira, my dream for you is that you find your passion and that you continue to work to create a more just world. For my partner, Jamual, my love since I was fifteen, I have no words to thank you for everything you have always done for me and our family. The good things that I have been able to contribute to this world are because you have always supported and loved me. I love you!

**Alicia**. First and foremost, I want to thank the authors for writing these amazing *testimonios*. We are all incredibly grateful for trusting us with your stories and for taking the time to relive the many times traumatic episodes you all share with us and allowing us to witness and celebrate your many successes. I also want to thank *Mr.* for his love and unconditional support. *A mis padres. A mi abuelo Max, (fintas), quien siempre fue partidario de mi educación y maestro, en ejemplo, de disciplina y trabajo duro. A mi abuela Lupe, (Buki), que paso horas ayudandome a aprender las tablas de multiplicar. A mi abuela Hildelisa (Licho), que me amo tal como siempre he sido (su morisqueta la extraña todos los dias) Y finalmente a mis abuelos Agustín y Alicia, que nunca conocí mas cuya sangre corre en mis venas y alimenta mi trabajo y misión.*

**Luis**. *Para mi gente, que demuestran los valores del empeño, trabajo duro, y sacrificio. Para mi gente del pasado, que han sacrificado tanto para que yo pueda llegar a donde estoy hoy. Para mi primo Alejandro Sordo Vieira, el niño que me enseñó que uno puede sonreir puramente al frente de una tragedia. Para mi mamá y papá, las personas que siempre estarán a mi lado. Para Sarah y Joaquin—los amo con todo mi corazón. Con cariño para toda mi gente.*

**Vanessa**. This book would not be possible without the authors who shared their powerful stories of success, hardship, and perseverance. I am grateful for your trust and confidence in us as safe-keepers of your stories. Thank you for paving the way for many of us. To my fellow editors, thank you for your dedication and work to make these stories shine and reflect the authors' authentic voices. *Para mis padres, Gloria y Rafael, y mis hermanos, Glorimar y Rafael José, gracias por siempre apoyarme, ser mis fans número uno y una parte vital de mi historia. Para mis abuelos, que aunque su vida no fue fácil, se empeñaron en crear un mejor futuro para nosotros. Su legado vive a través de nosotros—los amo.* To my partner Gert, for your love and support, and for always empowering me to use my voice to make a difference. To Elke, for being my friend, cheerleader, and walking by my side in my journey. *Para mi gente, que estas historias sean una celebración de nuestro recorrido y el camino que queda por recorrer.* To my friends that have become like family, you inspire me to fight the good fight. Finally, to the reader, thank you for honoring these stories with us.

**Rosaura**. *Para mis padres, Rosa y Victor, que han sacrificado tanto por mi y siempre me han brindado amor y apoyo incondicional. Me han enseñado lo importante que es luchar por lo que quieres y nunca darte por vencido. Mi padre siempre dice "No me digas que no puedes, dime que no sabes." Toda mi vida he llevado esta frase en mente que me recuerda que yo puedo, aunque a veces las cosas sean difíciles. Los amo con todo mi ser.* For my little sister Fernanda, who is my partner in crime and has always been unafraid to stand up for what she believes is right and just. She inspires me to do better and be better every day. I love you sis. *Para mi tío Rafa, que ya no está con nosotros, pero con su ejemplo me enseñó a nunca dejar de aprender. Te extraño. Para toda mi maravillosa familia, son mi fuente de inspiración. Los quiero. Y finalmente, para mi esposo* Alek, who is my biggest supporter *y una de las personas con el corazón mas grande y lindo que conozco. Siempre ha estado a mi lado ayudandome y apoyandome en todo momento. Te amo hoy y siempre,* my love.

**Andrés**. *Para mis creadores, Mami (Sara) y Papi (Rodolfo). Para mis hermanas, Alejandra y Sarah. Para mis padrinos, Ana y Julio. Para mi familia con la que el universo me bendició (mis tíos, tías, primos, y sobrinos) y para mis amigos que se han convertido en familia. Gracias por su amor y por apoyar mis metas. Para mis maestros que me enseñaron a apreciar la belleza de la matemática. Gracias a todos que han cruzado mi camino y me han ayudado a redefinir lo que significa ser matemático. Gracias a la vida que me ha dado tanto. ¡Pura vida!*

For my creators, Mami (Sara) and Papi (Rodolfo). For my sisters, Alejandra and Sarah. For my godparents, Ana and Julio. For my family with whom the universe blessed me with (my uncles, aunts, cousins, and nephews) and for my friends who have become family. Thank you for your love and for supporting my goals. For my teachers who taught me to appreciate the beauty of math. Thank you to everyone who has crossed paths with me and has helped me redefine what it means to be a mathematician. Thank you to this life, which has given me so much. *¡Pura vida!*

# Dr. James A. M. Álvarez

## Early Life

**Olga's story**. My mother, Olga Mendoza, grew up in a small community formed around the turn of the twentieth century as a "company town." The company produced large clay sewer pipes in the town of Saspamco, Texas, which is actually an acronym for San Antonio Sewage Pipe and Manufacturing Company. Before my grandfather, Federico Mendoza, was murdered in 1939, he worked at the company, and so did several of my great uncles and later so did my brothers. This town was (and still is) comprised of more than 95% Hispanic people of different backgrounds. That is, some had been in the area when Texas was part of Spain and Mexico, while others (like my great-grandparents) had come to the area in the 1880s and 1890s, and the remainder were those who had left Mexico during the Mexican revolution in the 1910s.

Dr. James A. M. Álvarez

Illustration created by Ana Valle.

My mom went to school in Saspamco through the eighth grade. However, high school grades in Saspamco were not accredited by the state of Texas. So students who wanted to earn a valid high school diploma had to go ten miles away to attend school in the nearby town of Floresville. During this time of the Great Depression, Maria (Mary) DeAnda, my widowed grandmother, raised poultry and milked cows to support my mother and her brother. This brought enough money to get by, but not enough to afford to pay someone to give Mom a ride to get to high school every day. So at the age of 12, my mom moved to live with her aunt and uncle in South Texas so that she could attend ninth grade at an accredited high school.

The next year, she returned to the farm near Saspamco because Grandma had found a neighbor in Saspamco who worked in Floresville who agreed to charge her $3 per week to take mom to and from school. However, the $3 a week charge only paid for transportation for Monday through Thursday and Grandma could not afford the additional cost for transportation on Friday. Luckily, Ms. Wiseman, a caring teacher, noticed that my mom was an excellent student, but that she missed school every Friday. Asking her about this and learning of the financial hardship to attend school, Ms. Wiseman generously worked out a plan with my grandma to have Mom stay in town with her each week. This way, Mom could walk to school daily rather than missing one school day per week. Ms. Wiseman provided these accommodations at no cost to my family.

Generous people, like Ms. Wiseman, made it possible for my mom to continue her education; an education which proved to benefit future generations, including me. Yet other people, like Mom's Algebra I teacher, acted in an opposite manner. Mom recalls being the only Mexican-American in the class and also the only student who did not receive a textbook from him! Needless to say, my mom did not have a great mathematics background.

Sadly, such discriminatory experiences were not isolated incidents and they affected many within our family. Mom was not allowed into the pool at Landa Park in New Braunfels, Texas, on her high school senior trip because Mexicans were not allowed in the pool; my grandfather was fired from a job in Sugarland, Texas, when the foreman realized he was Mexican-American; as the best students in their Spanish class at Texas A & I College (now Texas A & M Kingsville), Mom and another classmate were invited by their professor to eat at the famous Kings Inn and the professor had to ask special permission for Mom to enter because Mexicans were not allowed in; professors at different institutions would routinely give C's to my mom, her cousins, and other Hispanic students regardless of their academic performance.

Being discriminated against for having an accent, for having brown skin, and for being economically disadvantaged, shaped the way my mom decided to raise my siblings and me. This in turn greatly affected our drive to strive for excellence as a way to mitigate expected, or unexpected, discriminatory practices that would surely come our way.

In 1955, Mom returned to the community where she grew up as a teacher for Saspamco Independent School District. During this time, Mom had six children (see photo below).

**James' early years**. As the fourth child of six, I vividly recall living in the two-bedroom teacher's cottage. While the cottage was very affordable, it was not the best living situation for a family of more than seven, especially as it did not have running water in the kitchen.

Picture of my grandma with my siblings and me 1970.

Photo courtesy of James Álvarez.

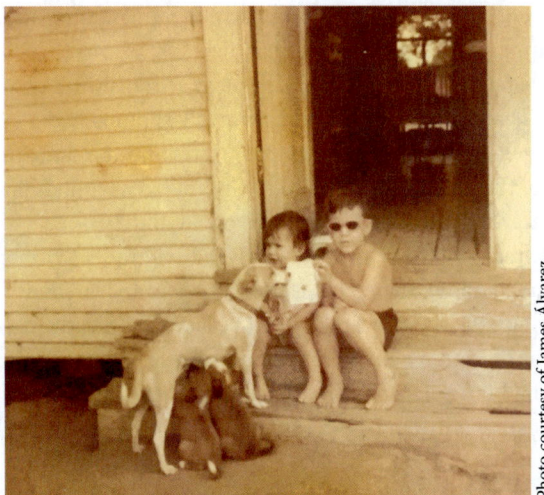

Photo courtesy of James Álvarez.

Picture of my little sister Olga (who was appointed judge by the governor of California in 2019) and me at the steps of the teacher's cottage.

Whatever challenges this living arrangement posed, I grew up on the idea of "*la ropa sucia se lava en casa,*" which translates to "not airing one's dirty laundry in public." This motto was a way to protect ourselves from being left out given our financial situation and it led to my being very reserved as a child. I did not share personal details with others which, in turn, gave my schoolmates the impression that I was just like everyone else in my class and that I was having the same socioeconomic experiences as them, even if it was not so.

Nonetheless, by the late 1960s, while being at a rural school with only four grades, I was able to freely wander from the cottage to see Mom during the day. Self-identifying as a "momma's boy," I spent a lot of time in my mom's first-grade classroom (see below).

I cannot remember how I learned to read or add or when I learned about numbers, but I remember being good at it well before I began first grade. In fact, when I was only four or five years old, I would sit at the back of Mom's classroom and follow her math lessons

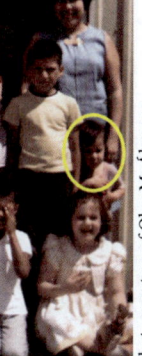

Photo courtesy of Olga M. Álvarez.

My mother's class in 1968 with me at her skirt.

along with all of the other students. From an early age, I would tutor other students, assisting them with their reading and mathematics work.

By the time I was ready for first grade, Saspamco schools were closed and consolidated with the schools in nearby Floresville. At that time, Saspamco had a population of about 200, while Floresville's population was over 3,000. Around the same time period, my family moved out of the teacher's cottage to live in my great-grandfather's house, which my dad, Jonás Álvarez, had remodeled with his own hands.

I first remember realizing I was good at math in first grade. My mental math skills were very sharp and I had already learned most of what we were learning. This meant that, since the school grouped classes by ability, I was placed in the advanced group. I may not have been conscious of it on my own, but Mom made sure that I understood the responsibility that came from being placed in the advanced group. In a time when my school was about 50–60% Hispanic, only about five of thirty students in the advanced group were Hispanic.

I later learned that several of the white students in my advanced group were placed there because of their parents' desire to keep them out of classes with large percentages of Hispanics, rather than based on their academic potential and achievement. Mom, a second-grade teacher at my school, witnessed the ability grouping process in which these requests were routinely made. Not only this, but she also was subjected to some white parents not wanting their children to be in her classroom because she was Mexican; at that time, there were only a handful of Hispanic teachers in the entire school district.

I distinctly remember being made to feel like an outsider by some of my classmates. To try to fit in, I continued to be quiet and reserved, but I have never forgotten a classmate telling me that my father was a "Meskin lover" because Mom was "Meskin" (a derogatory term for Mexican). Not being white, I could not be part of his group of friends. Remembering my mom's mantra that "education and knowledge" were pathways out of poverty and the only way to combat ignorance, I tried to look past such hurtful statements and focus on learning and being the best student I could be. Not only this, but from my earliest memories, I wanted to be a teacher so that I could help others, just as my mom did. Thus, throughout my childhood, I always felt that I needed to work hard and learn as much as I could; not only for myself, but also for my community.

Always enjoying mathematics, I was fast at arithmetic computations. Ms. Tipton, my third-grade math teacher, would give us fifty arithmetic computation problems, including exercises such as multiplying and dividing four-digit numbers by four-digit numbers, to work each day before starting class. I enjoyed always being the first or second student to finish the computations, and once completing them, I took the opportunity to help others finish their work. One of my earliest joyful moments in my early mathematics education was in fifth grade, when Ms. Higgins taught us about other bases. I was fascinated by working in other bases and I can still recall a paper she handed back to me in which she complimented my work when working in base 5.

In sixth grade, in addition to reading and numbers, I developed a love for music as I began learning both the piano and trumpet. I enjoyed the piano from the beginning, but not so much the trumpet. I struggled at first, but persisted for two reasons: first, my parents had sacrificed a lot to buy me the trumpet, and, second, because my mother always told me "don't be a quitter." The persistence paid off and I played a solo at the sixth-grade

Photo courtesy of James Álvarez.

Performing my senior solo in high school (May 1984).

concert and was praised by my band director for being the only student who had completed a music theory assignment that would have been challenging for students in a college-level music theory class. In fact, I continued to be a soloist and participated in many honor bands throughout high school. This extended to participating one year in college marching band as well as a few years in the UT-Austin Mariachi *Paredes de Tejastitlán* when in graduate school. Also, in high school I was recruited to play the organ at our little church in Saspamco. I found this quite stressful because I am a perfectionist and I disliked making mistakes in public. But, when you're from a small town, then you have to rise to the occasion.

As I moved to junior high, my older sister Mary started high school. Mary would often talk of her perceptions that only white students could be in certain classes and school clubs. Despite her good grades in junior high, the school counselor advised her to become a secretary and take high school classes that would lead to that—it is worth noting that many other Hispanic girls of that generation had also received the same advice by this counselor. Mary pushed back and demanded to be allowed to take classes that would allow her to become a teacher, as she also wanted to pursue a career in education like her mother. She won that fight, and I am happy to report that Mary is now Dr. Mary E. Carrasco. She holds an EdD and has taught school for over 30 years.

With my own meeting with the counselor looming, I made up my mind that I would not be subjected to the "othering" my older sister had experienced. As one of the best students in my grade, I was not going to let others limit me. So, I braced myself for the meeting with that same eighth-grade counselor. At the meeting, she looked at my grades and scores and said, much to my surprise, "you can be anything you want to be!" As encouraging as this encounter was for me, it made me reflect on how different my sister's experience had been and how back then and even today, the female experience in academia is filled with its own unique challenges.[1]

---

[1] In fact, as a female student, one of the editors of this book had a similar experience thirty years later, where her younger brother was encouraged to pursue higher education while she was advised to become a bilingual secretary.

Miss Escamilla, my only Hispanic mathematics teacher, taught me Algebra I. She was an amazing teacher who guided me in learning more than the planned curriculum so that I could prepare for a mathematics contest in algebra. This was a mathematics subject-based contest sponsored by the Alamo District Council of Teachers of Mathematics in San Antonio for schools in the San Antonio, Texas, region. I placed third in the contest; because we were a small school, we rarely placed at the contest and people took notice. I can still picture her working examples on the board and my excitement to show her my work on problems. Her encouraging and kind words had a profound impact on me. However, I quickly learned that some of my white classmates, when comparing me to high achievers in our school, gossiped that there was no way that I could be smarter than they were because I was "Meskin" and from Saspamco, and now we are at the point where the story repeats itself.

This was one of several discouraging and discriminating events I experienced while in high school, just as my mom experienced in her youth. One very troubling memory comes from my geometry class, where my teacher enforced a policy of not giving a final average grade of more than a 98%. So, even when my average was 100%, she would give me a 98% on my report card. Aside from the sheer injustice in such a grading scheme, my concern ran deeper. I knew that I could become class valedictorian, an important designation for scholarships to college. Placing a ceiling on my grade only advantaged other students for such a designation. My father called the teacher and asked her to give me the grade I earned. She said, "no, because nobody's perfect." I remember my father's response: "well, maybe he's perfect in your imperfection." My father, a former priest, was well versed in logic and philosophy, yet could not get my teacher to understand that a student's average grade in a course tells you more about a teacher's expectations than anything about perfection. Hence, the 98% grade in geometry remained.

These negative experiences did not end in geometry, as the advanced mathematics teacher routinely told us not to take calculus when we got to college. His reason was personal. He told us that he had done that and failed. One can only guess what little this teacher must have thought of our ability or future trajectories, or whether he understood that he was not really preparing us to be successful in a future calculus course.

Fortunately, neither of these experiences discouraged me. Instead, I worked hard in high school taking advanced classes and learned enough to earn ACT scores that translated to a year's worth of college credits when coupled with College Level Examination Program (CLEP) exams. Furthermore, I became valedictorian of my high school class which entitled me to free tuition at a public university in Texas, and I earned a Presidential Scholarship to East Texas State University (now Texas A & M Commerce), where I aced my calculus courses despite my high school teacher's discouraging counsel.

## Higher Education

**Undergraduate education**. While at East Texas State University, I approached Dr. Stuart Anderson inquiring about completing an honors thesis in mathematics. Dr. Anderson, a topologist, told me he was unsure of what kind of research I could do, as an undergraduate. This led me to work on an honors thesis in physics, my second major.

Picture of me in 1987 with Dr. Jerry Morris, President of ETSU, for having been named Outstanding Senior Physics student and for Academic Distinction.

After beginning the physics honors thesis, I realized that I did not like the particular project I was pursuing. I also calculated that if I did not complete an honors thesis, I could graduate with academic distinction after only completing six semesters. The latter of which would allow me to return closer to home; a great benefit, as I was incredibly homesick. In light of this, I dropped the honors thesis and instead completed independent study courses with Dr. Anderson, who offered these courses so that I could prepare for graduate school in mathematics.[2]

**Graduate education**. Having finished my undergraduate degree in three years, I entered graduate school at the University of Texas Austin. Yet, I had not taken an undergraduate real analysis course. Perceiving myself as a highly capable mathematics student, I felt terrible that I had to take undergraduate real analysis when I was a *graduate* student. My lack of understanding of graduate education in the sciences and not having any role models to assure me that this was okay, was disconcerting. I began to question whether I was really graduate school material, while not remembering that I was a 21 year old who had opted out of a fourth year of undergraduate studies. Yet, my choices and poor graduate advising made my first few years of graduate school emotionally difficult. For example, I was not advised against taking the second semester of graduate abstract algebra without having had the first semester of the course. Whether such advising was due to plain negligence, or potentially something more problematic, I will never know.

My graduate education was supported through a Graduate Opportunity Fellowship for underrepresented minorities. While it provided enough support so that I could be financially self-sufficient, I felt left out of the group of graduate students who were serving as teaching assistants. On several occasions my fellow classmates clearly equated my being on a Graduate Opportunity Fellowship as meaning that I was not qualified to be in graduate school. Yet, time would prove them wrong since of the approximate 30 students who

---

[2] Dr. Anderson told me later that he always regretted not being more proactive about finding me a project for a mathematics honors thesis. But, Dr. Anderson's Moore-Method topology course, which he offered on his own time to me and another classmate, provided a strong mathematical foundation for my future graduate studies.

entered the PhD program in my year, I estimate that only six of us eventually finished the degree.

Preparing for my preliminary exams at UT-Austin was quite stressful. Graduate students had access to copies of old preliminary exams along with recommended resources and books from which to study. However, there was an important book that I did not have and which had been indefinitely checked out of the library. As is still commonplace, the book was too expensive to purchase on a graduate fellowship stipend. I decided to stop by Dr. Bill Beckner's office to discuss some preliminary exam problems and mentioned that I could not get a hold of this particular reference book. Dr. Beckner pulled it from his shelf and told me that I could use his copy, but requested that I not bend the pages—one of his pet peeves. I was very happy to use the book and felt grateful that Dr. Beckner trusted me with his book. This gratitude only grew given that after I passed my preliminary examination, Dr. Beckner told me that I could keep the book because he had received another copy from the publisher. This kindness, along with the fact that I was really enjoying analysis and probability theory, helped cement my desire to have him as my dissertation advisor.

During that time, UT-Austin began the Emerging Scholars Program (ESP) in calculus, which was motivated and informed by the research work of Dr. Uri Treisman. ESP targeted students coming from historically underrepresented minority groups in mathematics-based disciplines just like me. It focused on academically capable students who could benefit from experiences to close academic gaps and enrich their understanding of mathematics. At a spring departmental picnic, Jackie Bacon, a fellow graduate student, spoke to me about her experiences teaching in ESP that first year of its inception. Fascinated by her comments and learning of the goals of ESP I told her that I really wanted to teach in the program. I was eager to make a difference just like Dr. Treisman.

Working in ESP gave me the opportunity to develop my teaching and channel my creative energy. I worked with amazing students who came from similar backgrounds like mine. Two such students are Dr. Rey Rivera who is now President of Estrella Mountain Community College and Dr. Cristina Villalobos[3] who is now a Professor of Mathematics and Associate Dean at University of Texas Rio Grande Valley. My excitement in facilitating problem-solving sessions, seeing my students make deep connections in calculus tasks, and seeing them flourish, gave me great joy. I often thought about creating opportunities for students to understand mathematical concepts deeply—opportunities I felt I lacked in my own background, as much more could have been asked of me.

As I taught in ESP, I continued my dissertation research in percolation theory, an area of statistical physics and mathematics that describes the behavior of connected clusters in a random graph. Problems in percolation theory are very simple to state, but extremely difficult to prove. The independence required to conduct my research and the intensity of the research process left me feeling very isolated.

As a result, but perhaps not entirely, I became uninterested in my research and I began to dedicate more time to my ESP teaching. After just a year as a teaching assistant in the program, I was leading the annual instructor professional development workshop for faculty and graduate students who were running programs across the country. I also be-

---

[3] Dr. Cristina Villalobos is also featured in the book.

Picture with my parents during my PhD graduation in 1996.

gan giving presentations with Dr. Treisman about many aspects of ESP. For the second or third year of the UT-Austin training, I came up with types of mathematical problems and designed tasks for the worksheets used in the professional development of mathematics instructors.

Luckily, my PhD advisor allowed me to work on what I was interested in and gave me the independence to pursue my research while continuing my involvement with ESP. Yet, the continued feeling of isolation led me to realize that I was much more passionate about finding ways to provide deep mathematical learning experiences for students. This passion turned into a job. At the completion of my dissertation, Dr. Treisman offered me a post-doctoral fellowship at the Charles A. Dana Center for Mathematics and Science Education at UT-Austin.

A picture of me in 1998 with the ESP staff when they presented me a certificate at the end of my postdoc for all my service to ESP.

Jumping at the chance to learn more from Dr. Treisman, I immersed myself in mathematics education. A major accomplishment while at the Dana Center was my work as one of the primary authors on the mathematics chapters of the Dana Center's and College Board's Mathematics Vertical Teams Toolkit.[4] My work at the Dana Center also began my long involvement in various capacities with the Texas Essential Knowledge and Skills in Mathematics.

After completing my postdoctoral work I had a few job offers that would allow me to continue my research in percolation theory and mathematics education. But I decided that I could not split my commitment between mathematics research and mathematics education research. So, I accepted an attractive offer from the Department of Mathematics and Statistics at Texas Tech University. This position allowed me to take leadership in the Master of Arts program for teachers, offer graduate student assistant training, and conduct research in undergraduate mathematics education. This new opportunity led to my decision to become a mathematics education researcher.

## Current Life

**Professional career**. Due to my abrupt change in direction after my PhD, my mathematical research was limited to the research I conducted for my dissertation. In this work, I examined percolation on a randomized fractal using techniques from percolation theory, fractal geometry, and probability theory. I established threshold probabilities for determining when paths in the fractal would only be a local or surface effect versus one that would allow percolation through the structure.

My mathematics education research has been varied. In my early work, I focused on building the Master of Arts in Mathematics program for secondary mathematics teachers, after I moved from Texas Tech to UT-Arlington, and much of my energy focused on garnering funding to support the program. This led to creating an important resource for mathematics teacher educators entitled, "Strengthening and Supporting Standards-based Mathematics Teacher Preparation." Later, capitalizing on my early work on the 1998 Mathematical Vertical Teams Toolkit, which was disseminated nationwide for many years by the College Board, I was a co-principal investigator on a $3 million GK–12 grant from the National Science Foundation (NSF). This grant allowed me to work with a high-needs school to develop graduate students' capacity to communicate and integrate their mathematics research into school curriculum. In partnership with teachers, this work vertically traced back key ideas and elements of our research to school-level mathematics where they were placed.

Almost simultaneously, I also received almost two million dollars in funding from the NSF as a co-principal investigator for the Arlington Undergraduate Research-based Achievement in STEM project. This project enabled creation of an Arlington Emerging Scholars Program in mathematics and chemistry whose focus was on broadening the reach of the more traditional Emerging Scholars Program model.

My more recent work involves mathematical problem solving and the development

---

[4] Advanced Placement Program, Mathematics Vertical Teams Toolkit. Epperson, James, Holtzman, D., May, S., Sandow, D., Stanley, D. New York, NY: "The College Board and The Charles A. Dana Center, The University of Texas at Austin".

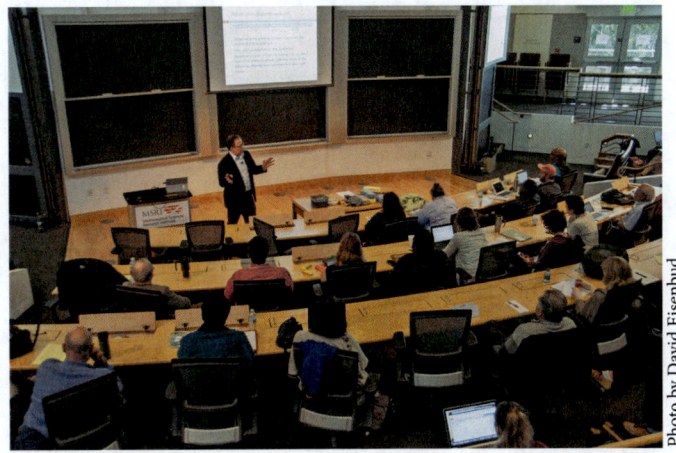

Presenting my work at the Mathematical Sciences Research Institute in 2017.

of mathematical knowledge for teaching. I have contributed as one of the lead authors of the Classroom Practices chapter of the Mathematical Association of America's (MAA) *Instructional Practices Guide.*[5] My work as a co-principal investigator on the NSF-funded Mathematical Education of Teachers as an Application of Undergraduate Mathematics (META Math) project is laying the foundation for finding ways to integrate applications to mathematics teaching and explicit connections to school mathematics into mainstream mathematics courses for prospective secondary mathematics teachers.

I am grateful that my research and teaching have not gone unnoticed. I have been recognized six times with the UT-Arlington Provost's Research Excellence Award for my research and scholarship contributions. In March of 2020, I was selected as UT-Arlington's nominee for the Piper Professor Award, a statewide award given to only 10 professors across the state of Texas. Each university selects one nominee to move forward to the statewide competition. I have also been recognized for my teaching at many levels. This recognition includes the University of Texas System Regent's Excellence in Teaching Award, which carried a $30,000 award, and an outstanding UT-Arlington Honors College Faculty Teaching Award in 2016. Because of my success as an educator, I carry the title of "Distinguished Teaching Professor" which is granted to faculty inducted into the Academy of Distinguished Teachers at UT-Arlington.

**Family**. As is often attributed to Latinx/Hispanic culture, my family is extremely important to me. I want to point out that my position at Texas Tech not only gave me a new professional opportunity, but in fact it is where I met my wife Dr. Minerva Cordero,[6] also a Latina mathematician. I love spending time with my family and traveling. Minerva and I love to dance, and we do so any chance we get. We now live in Arlington, Texas, close to my family and have raised two wonderful sons.

On a personal note, I enjoy music, singing, and watching international movies including Spanish-language *novelas* and mini-series. I am also an avid genealogist and I have

---

[5] Martha Abell, Linda Braddy, Doug Ensley, Lew Ludwig, Hortensia Soto-Johnson (Editors). *MAA Instructional Practices Guide* 2017, Washington, DC: Mathematical Association of America.

[6] Dr. Minerva Cordero is also featured in the book.

My wife Minerva and sons Alex and Nicholas.

traced my Mexican (e.g., Spanish, Mestizo, Mulato) ancestry in Mexico to the 1500s and further back to Spain to the 1300s. I believe my indigenous ancestry is mostly Chichimeca from San Luis Potosi, but I still have more work to do to verify this completely.

## A View Toward the Future

I believe the way for those outside the Latinx/Hispanic community to better support us is to strive to find common ground and aspirations that we can all agree upon.

*What's the difference between a Latinx mathematics major and a non-Latinx major?*

They may have been given very different messages about who can and can't do mathematics. It is our job to replace negative messages with authentic encouragement, support

Dr. James A. Mendoza Álvarez

positive messages with excellent teaching and mentoring, and build upon successes by providing opportunities for advancement.

*What's the same about a Latinx mathematics major and a non-Lantinx major?*

They both love mathematics, and they chose mathematics because they identify as a "math person." Use this common ground. All majors need support and guidance. All majors should be given opportunities to further their mathematical interests. It is in making these equitable where strategies need to diverge. Negative stereotypes, unproductive classroom interactions, and shifting priorities unrelated to mathematics require us to reexamine how to provide the right support for Latinx majors that will enable them to take advantage of the same rich mathematical experiences provided to others. We need to re-imagine ways that all students can flourish.

# 2

# Dr. Federico Ardila Mantilla

## De Donde Vengo Yo

My elders come from little *pueblitos* in the mountains of Santander, in northeast Colombia. My dad's family would be the ones to call if you wanted a serenade for someone in Zapatoca. My great-grandma would host the afterparty and challenge the young ones to trash-talking battles in rhyme. Meanwhile my great-grandpa slept on a hard wooden block, in contrition for his sins. My dad, Jorge, loved math and science as a kid in Bucaramanga and wanted to study engineering. Fortunately, his dad won the *Totogol*—a sort of lottery where you had to guess the scores of the national *fútbol* tournament—winning just enough

Dr. Federico Ardila

Illustration created by Ana Valle.

money to open up the corner store that supported the family for the next few years, and send my dad to college in Bogotá. My father began his career engineering sewage systems. With time, he became really interested in designing systems for groups of different people to work together, *en armonía*. He will tell you that he always prefers to avoid conflict; but if he has to, he'll always fight for the weaker side.

My mom's parents migrated to Bogotá escaping from circumstances I will probably never know, and finding common ground in that. My grandma would proudly say she was the first woman to wear pants in her town. She made sure that my mom never learned how to cook so she wouldn't end up oppressed by some man. My mom, Amparo, was part of the first generation of professional sociologists in Latin America; she managed to graduate before they shut down her department for being too revolutionary. I recently learned she almost called me Fedor, the Russian name for Federico. Amparo devoted much of her life to the NGO[1] she founded and directed her whole life. She helped institute a national sex-education program—this included teaching at my school, and my friends' schools, and even on TV sometimes; I was not amused at the time. She organized women to battle domestic and sexual violence. She helped over-policed neighborhoods develop conflict

---

[1] NGO stands for Non-Governmental Organization, which is a nonprofit organization that operates independently of any government, typically one whose purpose is to address a social or political issue.

Photos courtesy of Federico Ardila.

My mom and dad showing us the way.

resolution protocols, away from the police. She fought the National Rifle Association at the United Nations, as she tried to stop the U.S. government from sending weapons to Colombia as part of the "War On Drugs". She seemed to think there was no problem too big for her to tackle, no fight that she could not win.

From an early age I was exposed to the beauty of math and science, and I was instilled with a strong sense of social responsibility. One of the big questions of my life is to understand how these two interests and commitments can support each other. I have learned a lot from struggling to find answers to this question.

## My Early Years

I grew up in Bogotá, Colombia. I was a really shy and sensitive kid, excited about all sorts of things: *fútbol*, music, art, typography, reading, writing, math, among many others. My earliest math story, according to my dad, comes from when I was learning how to add. To pass the time while stuck in traffic, he asked me to add the numbers from 1 to 9. I told him "1+9 is 10, 2+8 is 10, 3+7 is 10, 4+6 is 10, and then 5, so the answer is 45." As a math educator, I can see now that my dad had the taste for a good math problem. My mom's parents challenged me to weekly games of Yahtzee and blackjack, for years. My math development was supported from an early age.

My school, *el Colegio San Carlos*, was the service project of a North Dakotan abbey. It was highly selective academically, but Father Francis, the school principal, insisted that the school should be affordable. In elementary school we had some awfully conservative North Dakotan teachers, who punished us physically, and implicitly taught us that American whiteness was something we should aspire to. In high school we had some pretty radical teachers, who argued about which guerrilla was most beneficial for the liberation of Colombian people. So that we might fully appreciate Baudelaire's poetry, our Spanish teacher had us act out a ritual with blindfolds and candles in the school theater, while we read *Las Letanías de Satán* from *Las Flores del Mal*.[2] Our teachers were given a lot of freedom, clearly.

---

[2] English: *The Flowers of Evil* is a volume of French poetry by Charles Baudelaire. First published in 1857, it was important in the symbolist and modernist movements. The poems deal with themes relating to decadence and eroticism.

My school had a really obnoxious motto: "The best for the best." In English, too! Even more obnoxiously, it seemed to work. Many of my former classmates excel in many fields today: math, biology, music, literature, film, dance, finance, medicine, and others. Surely this is in part due to the quality of the educational experience we were offered. But outside of school, most of us already came from relatively privileged situations. Inside of school, we were raised to think we were the best, and we deserved the best. Privileges, confidence, and entitlement go a long way.

I was raised to not misinterpret all of this as our being exceptional. I remember my mom returning, with a big smile on her face, from a parents meeting where she had a big fight with one of my teachers. She asked the school to stop telling us that we were better than the rest of the country: we were just hard-working kids with resources and opportunities. My teacher told her that I would be ashamed of her if I could hear her. I was really proud of her; I still am.

## Olimpiadas

When I was eight or nine, a test of the *Olimpiadas Colombianas de Matemáticas* arrived at my school. I still remember one early question: A snail climbs up a ten-meter pole. Each day it climbs up three meters, gets tired, falls asleep, and slides down one meter overnight. How many days does it take for the snail to reach the top? I was really excited and amused when I solved this problem. I had no idea that math could be this creative, even funny!

I fell in love with math through the Olympiads. They also opened up the world for me: I qualified for the Colombian national team, and got to travel to Sweden, Russia, Turkey, Hong Kong, Argentina, and Mexico as a teenager. These were amazing experiences mathematically and personally, and they showed me that this path could open doors for me that I had not imagined. It took me very long to understand and accept that I wanted to be a mathematician, but the seeds were planted early on. I have to thank the director, Mary Falk de Losada, who had a huge impact on my generation of mathematicians in Colombia. To tell the complete story, though, my older sister Natalia and cousin Ana María had done really well in the Math Olympiads before me. My sister tells me that the (very male-dominated) summer camps—the same ones that were so encouraging and nurturing to me— were really uncomfortable for her, so she left. It is much easier to be in the majority.

*Olimpiadas de Matemáticas* with Mary Falk de Losada and my cousin Ana María.

# Access

The Colombian market was closed to foreign products in the late 1980s and early 1990s. To keep up with my obsessions I had to be resourceful. *Fútbol* was not shown on TV; the closest I could find was an instructional booklet featuring Pelé. Each page had a photo of him doing a different move: "*remate con el borde externo*," "*cabezazo lateral*," "*parada de pecho*." I lined up all the neighborhood kids and made everyone do daily drills, one move per day, until we finished the book. We all have solid technique today. Apparently I've had a teacherly instinct from a young age.

My quiet rebelliousness drew me to the loudest music I could find. A proud teenage memory was getting *El Tiempo*, the largest newspaper in Colombia, to talk about Suicidal Tendencies, a largely Chicano hardcore punk band from L.A. that I loved. All of this music was completely unavailable in the Colombian market, so I plugged into an intricate system of trading tapes with kids I never met. The arrival of a mystery music tape—complete with our best rendering of the band's logo, which we had never seen, and our best understanding of the lyrics—gave me a thrill for days! I am still really thankful to "M.", a frequent correspondent whom I never met, for shaping my musical taste.

Good math books? I had no idea where to find them. Our teachers withheld their books from us, since that is where our exams came from. I heard someone had sneaked in some great little math pamphlets from Cuba and was selling them in the black market neighborhood where I bought pirate music tapes. They only had a few books but they were written by top Soviet mathematicians, and funded by the USSR, so they were excellent and very affordable. I first heard about the 4-dimensional cube from *El método de coordenadas* by Gelfand, Glagolieva, and Kirillov. It is still one of my favorite mathematical objects, and Gelfand's work has been hugely influential in my career.

At the International Mathematical Olympiads, I met young mathematicians all over the world and we agreed to send each other math problems that we liked. I loved receiving their letters and I spent many, many hours at home, trying to solve problems from my pen-pals, and from journals like *Crux Mathematicorum* that they sent to me. Meanwhile, I was one of those annoying kids who pretended that I wasn't even making an effort, that things just came naturally to me. I guess that growing up around lots of toxic masculinity, like I did in my school, I learned to do anything to avoid getting bullied. Being a "genius" was socially acceptable, but not being a hard worker! I still don't find it easy to struggle in front of other people, and that gets in my way as a mathematician.

# Psychology

The International Mathematics Olympiads made me confront the human, emotional side of doing mathematics very early on, *a las malas*. It was made clear to me that I was expected to win Colombia's first gold medal at the IMO, and I worked super hard to fulfill that promise. I reached the point where I felt like I could solve any math problem you would give me.

I went from overconfident to terrified during the course of one long plane ride. I knew I was a strong mathematician for a Colombian, but as soon as I saw the Russians, Americans, Chinese, Romanians, and Iranians (Maryam Mirzakhani being one of them)

I was convinced, inside my body, that I couldn't possibly do as well as them. I did not sleep the night before the exam, and I froze when I received it; I simply could not think. I watched the Chinese student sitting next to me hard at work, and despaired. I put in one of the historically poorest performances by a Colombian in the IMO. So did most of my teammates. Today, I have no trouble understanding students who tell me they freeze in front of an exam.

During my last year in the Olympiads, it was decided that we would address these emotional, psychological barriers as a team. As part of our training for the IMO, we enrolled in a group meditation program. (I recently learned that the founder was José Silva, a Chicano self-taught parapsychologist from Laredo, TX—a controversial and fascinating figure.) We were skeptical; I must say it felt a bit like a cult to me. But we meditated daily and even did a team meditation right before walking into the IMO exam. We already had a reputation for being the weird team that was always singing loudly wherever we went, and now people thought we were even stranger. That was the only Olympiad exam where I felt that I showed what I was capable of.

I did not win that gold medal because I could not solve the combinatorics problem. I felt OK with that. I did learn that, to be the mathematician that I wanted to be, I needed to be mindful of the human side of doing mathematics. Now I'm convinced that the time and energy that I spend growing as a human being—feeding my spirit, confronting my flaws and my insecurities—also makes me a stronger mathematician.

## Applying to College

As much as I loved to learn, I had a problem with authority, like many high schoolers. I would solve math problems, but not the ones given for homework. I would read literature and talk philosophy in my spare time, but not the books I was assigned. My friend Juan Carlos and I had a deal with the school guard: we gave him a sandwich or a beer and he let us sneak out of school. We spent countless afternoons in seedy *billares*, with a bunch of men from all walks of life, together in solitude; billiards cue in one hand, beer in the other, words rarely spoken. I barely graduated from high school. I failed physics, and to be allowed to pass, I had to come up with a *recuperación*. I made a video showing off the most physics-defying *carambolas* that I knew how to do. The video did not showcase much of an understanding of physics, but the teacher was impressed. The hours and hours I spent playing billiards paid off.

When it came time to apply to college, one of my friends told me he wanted to go to this U.S. school that had strong science and a great financial aid policy: if you got in, they'd make it possible for you to attend. I had never even heard of this school, and my family certainly could not afford it. But this financial aid deal sounded good, and the application materials had an intriguing punk-rock aesthetic that spoke to me, so I applied. I was admitted and went to MIT.

Had I known how selective MIT was, I would have known that I wasn't going to get in: I was failing half of the subjects in school at the time! I probably wouldn't have bothered to apply. My ignorance helped me.

The accidental lesson I learned is that I think one should overreach for opportunities in life. There are many people, institutions, structures, who will close doors on us. Our job is to never close doors on ourselves, but to try to open them, push hard to try to get in, and if we do, make sure to keep those doors open for others to come in.

## Undergraduate Years

At 17, migrating to a new country and gaining access to an MIT education was tremendously exciting, if a bit scary. It was also a big culture shock. The undergraduate experience there was brutal. When I saw the amount of work that was assigned, I honestly thought professors were not being serious. I also played soccer for MIT, so the daily practices and games made it impossible to complete the work. That was fine by me, I was (maybe too) used to not completing work. But almost all my classmates had always been at the top of their class, had never failed an exam in their lives, and at MIT, many of them failed. This made for a horribly unhealthy environment: people worked way too hard, drank way too much coffee, and pulled all-nighters every other night. This didn't work for me; if I was going to succeed here, I would have to do it my way. Of course, I did study harder than I ever had. But I moved off campus and tried to have a balanced life outside of school.

The flip side, which I only realized many years later, is that I was extremely isolated as a mathematician. I felt so different from my classmates, and found the social atmosphere so uncomfortable that I avoided the department altogether. During my whole undergraduate career, I never collaborated with anyone on homework or went to a single office hour. I am not proud of this and certainly don't recommend it. I just thought that's how it was. But now I know that many students from underrepresented groups experience a similar isolation. I often find myself advising students not to let this happen: find people you trust and talk math with them; get used to the discomfort of discussing things you don't quite understand yet and grow from that discussion. Don't let your lack of confidence or your pride, two sides of the same coin, get in the way of your learning. But that's easier said than done, and really, this isolation is a big systemic problem that our math departments and programs need to face.

Aside from my classes, I was still obsessed with problem solving and spent probably too many hours preparing for the Putnam competition. I barely missed being a Putnam Fellow because I could not solve the combinatorics problem, once again. The mistake I made there, and I had also made in my last IMO, gave me a very helpful lesson in my research: you should not spend all your time trying to prove something that is not true.

I still didn't know what I wanted to do with my life until my senior year, when I took Algebraic Combinatorics with Sergey Fomin and Enumerative Combinatorics with Richard Stanley; I just loved them! Although combinatorics was always the hardest field of mathematics for me, these classes convinced me that I wanted to go to graduate school in combinatorics.

## Graduate Years

When it came time to apply for graduate school, I casually asked Richard Stanley for some good places to study combinatorics. He told me the names of four researchers, so I applied to those four schools, and to MIT. When it came time to choose schools, people asked me:

With Prof. Gian-Carlo Rota and his students at MIT.

"Wait, why didn't you apply to the top schools?" The thought did not even occur to me. I still didn't know what the "top schools" were, or why it was so important to go to them. (I still don't think the "top schools" necessarily offer the best education, but I can't deny that most math departments only seem to want to hire graduates of the same few schools.) So people advised me, "Well, given your options, you should stay at MIT." I did.

I took some great classes that shaped my mathematical interests: matroids with Gian-Carlo Rota, hyperplane arrangements with Richard Stanley, Coxeter groups with Sara Billey and Anne Schilling, algebraic geometry from Ravi Vakil, and Lie algebras from David Vogan. I took some awful classes too.

My PhD advisor, Richard Stanley, was and continues to be a huge mathematical influence to me. He is one of the founders of my field of research; he was able to draw deep connections between combinatorics and other fields that no one had foreseen before. More plainly, I really enjoy his taste in mathematics. At the time, Richard had 11 PhD students, and I only talked to him once or twice a semester. I didn't know this was unusual. He suggested some interesting problems, but I struggled to make progress.

I sometimes joke that I did my PhD in fútbol and salsa. I was on the MIT graduate soccer team, and at one point I was in eight other teams; so I played almost every night. I would also go dancing three or four nights a week with a crew of Colombians and my best friend May-Li, who became my life partner. Those were really fun and happy years.

(L) Salsa with May-Li. (R) *Fútbol* with the International Students Association.

And yes, it was fun, but years later I realized it was also escapism. Sometimes I just lost motivation and interest in my research; at an especially bad time, I hid from the math department for a whole semester and did construction work instead. I don't think anyone in the department noticed. I moved into a depressing little basement space, more boiler room than bedroom, where someone had recently died from alcohol poisoning. I was offered free rent in exchange for converting it into a livable space: I took out the flooring, put in a new floor and carpet, installed plaster walls and electricity jacks, hung a drop ceiling, the whole thing. I am not a very handy person; this was a real challenge. But I could see the progress! I would put in the hours and have something tangible to show for it at the end of the day, not like math.

Eventually I got back on track. I ended up finding my own research directions that I got very excited about. At each meeting I would give Richard a short update, and he would quietly walk to his massive book shelf, grab what seemed like a randomly chosen paper to me, and tell me to read it. It would take me some time to understand why this paper was just what I needed to read. I proved some theorems that I was proud of and people in my field found interesting, and I graduated.

## Postdoc Years

I turned in my thesis in December, applied for jobs in the U.S., and moved to Colombia for the spring semester, while job offers hopefully arrived. I soon learned that, upon leaving the U.S., my student visa was revoked and I was not allowed to apply for another visa for another 1–2 years. I had to cancel seminar talks, an invitation to the Clay Math Institute, and my plans to go to my own graduation. A few weeks later, I got a phone call inviting me to interview for a position at Microsoft. Now, maybe I shouldn't confess this, but I will: I knew nothing about this position, and I really had no interest in working at Microsoft, but I thought "Hmm… maybe Microsoft can get me a visa!" So I told them I would love to come for an interview. Two days later I had my visa. I could go to my graduation!

Of course, I went to the interview, and I was pleasantly surprised: I would spend one year at MSRI in Berkeley and one year at the Microsoft Theory Group—an academic research job with a really stellar group of mathematicians. Also importantly for me, one year of a corporate salary would allow me to pay all my loans within a year and move back to Colombia right afterwards. I took the job.

On a personal level, moving to the Bay Area was life changing for May-Li and me. It was the first time we felt at home in the U.S.: For a young Asian-Latinx couple fascinated with the countercultures of the world, tired of the Boston weather, excited to be reminded what fruit is supposed to taste like, the East Bay felt like an immediate fit.

I spent most of my postdoctoral years working at the intersection of combinatorics and algebraic geometry, with the guidance of Bernd Sturmfels and Sara Billey. I wrote my first joint papers with Carly Klivans and Lauren Williams, on the combinatorial foundations of tropical geometry; this turned out to be influential work. More importantly for me, I really enjoyed working on this topic and working with Carly and Lauren: they completely changed my understanding of how I like to do math research. I've only written a handful of solo papers since then.

## SFSU

At the end of my postdoc, May-Li and I were ready to move to Colombia—where, newly married, we were excited to build a life together—so we decided to only apply to a handful of dream jobs in the Bay Area, New York, and Vancouver. I was told that I was doing it all wrong and damaging my career. But we wanted a good life in a big, diverse city, with abundant opportunities for May-Li. I didn't want to pretend otherwise or play academic strategy games to try to get there. I got an offer at San Francisco State University, where I have been since then. Moving away from the Research 1 universities, I felt an unexpected, immediate sense of relief: I could leave the research rat race and do the work that I wanted to do, without pressure. Under these conditions, I feel I have done much better work.

SFSU has been a wonderful mathematical home for me. I feel lucky to be in a department with a strong combinatorics research group and amazing teachers and math education scholars to learn from, where service outside of mathematics is valued. The university has the attitude that our job is not to select and support only the top students, but to serve the students that come to us. I had never experienced this approach before in my schooling.

The first Ethnic Studies departments in the U.S. were established at SFSU in 1968, thanks to student activism, and that commitment to racial and social justice is still very much present on some parts of our campus. This has led to a wonderfully diverse community of students, whom I learn constantly from. I don't think I understood how much this would mean to me until I arrived to those classrooms; I felt so much more at home!

As soon as I realized that I wasn't returning to Colombia in the near future, I looked for ways to remain involved there. I started the SFSU-Colombia Combinatorics Initiative[3] and

The Encuentro Colombiano de Combinatoria in Medellín, 2016.

Photo courtesy of Federico Ardila.

---

[3] Ardila-Mantilla, Federico. "*Todos cuentan*: Cultivating diversity in combinatorics." *Notices of the AMS* 63, no. 10 (2016).

the *Encuentro Colombiano de Combinatoria*.[4] I am really proud of what we are building over the years. With the help of a strong international community, these have become really empowering mathematical spaces to many people, and led to fertile mathematical collaborations between Colombia, California, and many other locations. Within its means, SFSU has been tremendously supportive of this work.

There is a very skewed point of view in many math PhD programs, where people think the only successful route for a mathematician is to get a tenure-track job in a research school. If you can't do that, then you go make a lot of money in industry. Or if you're one of these unusual humanist types, you try to go to a fancy liberal arts school in a perfectly manicured little town. It felt to me—sometimes it still feels to me—that my research peers could not understand why I chose to go to a big urban public school with no PhD program, what strange value system led to this. What can I say? I really enjoy working here, and I really like the way that it has shaped me as a scholar.

## Balancing with Personal Life

Soon after moving to the Bay Area, I joined Left Wing Fútbol Club, an organization conceived as an alternative to the largely cis male, straight, white, affluent, win-at-any-cost, dominant culture of U.S. soccer. The team was made up of community organizers, educators, artists, rappers, and others trying to create political change through our professions. We had complete newcomers to *fútbol* and former professional *futbolistas*. With a good dose of Bay Area idealism, *fútbol* was seen as a metaphor for the kind of world we'd like to live in. Less experienced players would receive the ball most often, to have the opportunity to improve. In tournament games, everyone got equal playing time. The score was always 2-2. This will all sound very strange to anyone who played competitively before (like me!); it certainly wasn't a recipe for winning competitions. But on a good Sunday, it was the most liberating *fútbol* space I participated in in my adult life. I don't play actively with them these days, but LWFC showed me how to radically reimagine a space that I thought I understood very well.

My partner May-Li and I also started the DJ collective *La Pelanga* with some dear friends. We wanted a positive dance floor focused on the music and dancing, free of the sleaziness. We are obsessed with how music constantly migrates back and forth between Africa, the Caribbean, and the Americas, with no regard for borders. Recently we have started doing more cultural and political work. We DJed *Radio Ambulante's*[5] first live show and the Oakland Museum's exhibit on migration. We set the musical stage for a collective smashing of a *piñata* rendering of the border wall, led by artist Sita Kuratomi Bhaumik. We provided the soundtrack for the People's Kitchen Collective series of meals from the Farm to the Kitchen to the Table to the Streets, and for their annual celebration of the Black Panthers' Free Breakfast Program.

With time, DJing has become a second career for me. For many years I thought I should hide this from my colleagues and students. I couldn't really tell my department why I would rather not teach morning classes. I would arrive to class, still star-struck,

---

[4] Ardila-Mantilla, Federico, and Carolina Benedetti. "*Todxs cuentan* in ECCO: community and belonging in mathematics." arXiv preprint arXiv:2008.02877 (2020).

[5] *Radio Ambulante* is a narrative podcast that tells uniquely Latin American stories in Spanish.

Photos courtesy of Sana Javeri Kadri.

*Estamos Contra el Muro*, led by Sita Kuratomi Bhaumik, 2016.

wishing I could tell my students "You're not gonna believe it, I got to DJ for Celso Piña last night!!!" But one time a young woman danced over to the DJ booth to tell me that she loved the music; then I saw she was my student and she saw I was her professor; I don't know who was more in shock. Eventually the word got out, and I cannot tell you how many young mathematicians I have met who are also DJs, or bassists, or community organizers, or dancers, or athletes; they are usually worried that they will have to drop that to stay in math. So I've come to realize that it's important to share these parts of my life as well: for myself and for so many others who have felt that they need to hide the things we love to be accepted as "legitimate" mathematicians. Perhaps unsurprisingly, when I look at the way that I do my work as a mathematician in my research, my teaching, and my organizing activities, I find a very clear influence of the lessons I have learned from these spaces and these wonderful collaborators and friends.

## Balancing Act: Research/Teaching/Service

A constant source of joy and anxiety throughout my career has been to find the balance in my growth as a researcher, as a teacher, and as an organizer. I have never wanted to do one without the others. I am reminded of the end of my sabbatical year in 2013, when I had my head buried deep into lots of exciting research projects.

I was trying to figure out how I would balance them with my recently increased 3-3 teaching load—a strange perk of earning tenure in my department. Then, during the same week, I received invitations to become editor-in-chief of my favorite combinatorics journal (JCTA) and to co-direct my favorite program for young mathematicians of color (MSRI-UP). Each was a really exciting opportunity and a really overwhelming commitment. It was immediately clear that if I was going to accept one, I was going to accept the other also. I did.

My mathematical work is what gives depth and excitement to my teaching and community work. Ironically, my research is also what has given me the access to do the more human side of my work: whether I like it or not, in mathematics we are measured by our theorems, first and foremost. My teaching and community work are what give humanity and meaning to my mathematical work.

The classroom and mentoring my 50 thesis students has been an immense joy.[6] My so

---

[6] Ardila-Mantilla, Federico. "*Todxs cuentan*: building community and welcoming humanity from the first day of class." arXiv preprint arXiv:2008.02835 (2020).

Annual Federico-cooks-for-his-thesis-students lunch.

called "service work" cannot be seen as service work: I have grown and learned immensely from those whom I have supposedly "served".

Finding the time and energy to do all of these things is a different, challenging question. It certainly helps to be in a balanced relationship with my partner May-Li, who is tremendously passionate about her own career, who supports mine, and who constantly inspires me and teaches me. It certainly helps that we decided not to have children. It certainly helps that we don't have a TV. But I do know that I have had many times in my life when I am putting too much time and energy into my work. I have struggled with my health more than once because of it.

I will never forget the elder whom I met at a party a few years ago, who told me about her experience doing organizing work in Oakland and at SFSU in the 1960s and 1970s as a Black Panther. She asked me what I do and patiently listened to me go on for a bit too long about the many projects I was excitedly involved in. She lovingly put her hand on my shoulder, looked me in the eye, and said "Honey, it's wonderful that you are doing meaningful work; I can see that your students are lucky to have you. But you gotta take care of yourself. This work is a marathon, not a sprint. Can't forget to have fun while you're doing it, too." I'm working on that.

## The Difficulties

Still today, 25 years into my career, I sense in myself a tendency to make it seem like my path has been effortless, like I have it all figured out. I already said that the path that I walked on was paved by others, and it was already relatively smooth by the time I walked it. But Colombians frequently rank as one of the happiest peoples in the world, despite living through a civil war that has lasted my whole lifetime. We have an astounding capacity to rewrite our stories to make them a bit easier to swallow. And within mathematics, I've been hesitant to talk about things that have brought me a lot of anger and pain.

The numerous times my students and I carried 200 large bags of Colombian coffee to gift to professors at the International Math Olympiads, only to have some of them joke that we forgot to bring them the cocaine too. The time as a student when I proudly discovered a new mathematical object, simultaneously to some experts in the field, and one of them tried to get me to confess that I had stolen their work. The first time I was invited to give a talk in an algebraic geometry conference, and the experts spent 3/4 of my allotted time arguing about whether the question was even worth asking, and did not let me give my talk. The time that staff turned me away from the graduation ceremony of my Berkeley PhD student because they wouldn't believe me that I was his advisor. The time when an officer at the Spanish consulate refused to read my invitation and visa application to attend the International Congress of Mathematicians 2006 because she had too much work already, so I politely asked to speak to her manager, and she called the police on me, so I had to leave and miss the ICM. That Sunday when I forgot the keycard to enter my office building, and the SFSU police spent an hour confirming that my school ID and the departmental website were not lying, that I am indeed a math professor. The time I wrote a paper[7] on geometry, robots, and their use in society, and a referee wrote: "The math is very interesting; I don't know about these ethical ruminations. And is the author uncomfortable because a police robot killed a black person, or because they killed a person? What does 'Black Lives Matter' have to do with this?" The math conference where two of my students and I were kicked out of a bar, for no reason other than being the only non-white people there, cutting short our celebration of their recent accomplishments. The time when a book publisher read my book proposal and told me "Oh, I thought Colombians were good at drugs, not at mathematics." The time that a very prominent mathematician copied a large part of a grant proposal that I had shared with them, submitted it as their own, and won a large NSF grant with it. The time I received e-mail threats for talking about the role of mathematics in policing. Each time I did not make a big deal of it, as I had less power than them, or I felt this could hurt my career or get me in trouble, or I simply did not have the energy. Each time I told myself my work is not my life; this does not matter that much to me. But I still think about these episodes; they do matter.

I'll also state the obvious. Although I have received awards that celebrate my research, my teaching, and my mentorship, there are numerous times when I did not know how to support my students, mentees, and collaborators, and inadvertently caused a lot of pain. It often had to do with my inability to notice hidden power structures playing out in mathematical spaces that I was responsible for. I am extremely grateful to my students, my friends, my partner May-Li—mostly women of color—who have bravely and generously shown me what I could not see, and taught me to do a bit better every time. If there is something that I think I have done right in my career as an educator, it has been to listen to and learn from them.

---

[7] Ardila-Mantilla, Federico. "CAT(0) geometry, robots, and society." *Notices of the AMS* 67: 977–987.

## Our Worth

There are times when life simply will not allow you to function at work. If you look through my CV, you will find gaps in my "productivity" that have raised questions, that have closed some professional doors for me. Difficult times are hiding behind those gaps, but I am proud of them; there is a lot of growth there.

Several years ago, my mother suddenly suffered brain damage and went into a coma for the last nine months of her life. I spent a lot of that time with her, from near and from far; more than I had in 15 years. I learned to quiet down enough to accompany her very slow pace; to perceive her deep but very subtle communication. My mother had always been a proud and fiercely independent woman, constantly working at the service of others; depending this much on others was most certainly her worst nightmare. But these horribly difficult times also brought her moments of absolute bliss. Regardless of what she could or could not do at that point, Amparo was an immensely valuable human being, capable of giving and receiving love. This was a very hard lesson for me to learn, but it has been immensely valuable.

Soon after I learned that I had won the 2019 national teaching award of the Mathematical Association of America, I suffered a concussion on the *fútbol* field. I could not exercise, or read, or look at a screen, or listen to music, or talk to anyone for more than a few minutes. I do not mean to overdramatize the situation; but after several weeks of this, in a very real way, I was not sure whether I would still be a teacher, or a mathematician. I spent days and days and days in the woods and the ocean. This city kid, lucky to be living in the beautiful Bay Area, drew a lot of strength from the trees for the first time in his life. I learned how to meditate again and spent a lot of time doing it—there was little else I could do! This was a time of huge personal growth. So what if I was not a mathematician? Surely I was much more than a mathematician, right? (Right?)

I have always known that our profession (and our capitalist society, more widely) has the uncanny ability of making us feel that our worth is intimately tied to what we produce: the theorems we discover, the papers we write, the grants we receive, the awards we win, the classes we teach, the students we mentor, all of that. When we are not able to or do not wish to do any of these things, what is our inner worth? It can be a heavy question, but I think it is one worth considering.

## Embracing Nepantla

When straddling different worlds is a part of your identity, there is a lot of discomfort there, but there are also unique insights that you have to offer. I have been really inspired for years by the concept of *nepantla*, which I learned about from my friend Rochelle Gutiérrez.[8] As I understand it, *nepantla* is a Nahuatl word which roughly means *the space in between*. Chicana scholar and poet Gloria Anzaldúa used it to describe the point of view of living in the borderlands between different cultures; of not being in one place or the other; of seeing the world from different, sometimes contradictory lenses. It is a place of tension and discomfort, but it is also the place where transformation takes place. The

---

[8] Gutiérrez, Rochelle. "Embracing *Nepantla*: Rethinking 'knowledge' and its use in mathematics teaching." REDIMAT 1.1 (2012): 29–56.

discomfort is necessary and even desirable, as it signals the growth that is taking place inside you.

I find myself in this place very often: in my personal life, as a Colombian-American, as a Latinx-latinoamericano, as a member of a Chinese-Indonesian-Dutch-Canadian-Colombian-American household; in my research life, as an algebraic geometric topological combinatorialist; in my larger professional life, as a researcher-educator-organizer, as a somewhat public figure who often needs to hide from people, as a mathematician-DJ, as an insider-outsider who benefits from academic structures that he dislikes and is working to transform.

I imagine that many readers of this *testimonio* are very familiar with these in-between spaces. On either side of these visible or invisible borders, there are people and structures that do not understand where we are coming from, and may try to invalidate our point of view. But more interestingly and more importantly, there are bridges to be built. I think this is where we can best contribute. Building these bridges has been the most challenging and rewarding part of my personal and professional lives.

# 3

# Dr. Selenne Bañuelos

## My Parents

My parents grew up in very small towns in Jalisco, Mexico. My father was only able to complete the second grade of elementary school, just long enough for him to learn to read and write. He is the oldest in his family and worked with his father to help support his younger siblings. He began working in road construction at the age of fourteen. My father hung around the engineers so much that they started calling him nosey. He took whatever scraps of paper containing calculations that the engineers balled up and threw out. He taught himself geometry through those scraps of paper and still remembers discovering the Pythagorean Theorem in those notes. By the time he was seventeen, he bet an engineer that he could get a section of the road done with his own calculations. He did it beautifully.

Dr. Selenne Bañuelos

Illustration created by Ana Valle.

My mother's maternal grandmother lived in the nearby *pueblo*[1] of Mascota, Jalisco, and her paternal grandmother lived in the city of Guadalajara. My maternal grandfather was a *bracero*[2] in California since the 1950s. These circumstances allowed my mother and her siblings to study. My mother attended a boarding school in Mascota for elementary school and stayed with her grandmother in the *pueblo* for middle school. She then went to the city for high school, as there were no high schools in the *pueblo* at the time. She had hopes of becoming a schoolteacher but did not complete her schooling once she married my father. While in her forties, when her three children were all grown, she returned to school and earned a certification as a nurse's assistant. She has been working for the same clinic for more than a decade.

My father wanted to move to the U.S. to have better opportunities for work. He was in California in the late 1970s for a year before he returned to marry my mother. They spent their honeymoon avoiding border patrol agents as they crossed the Mexico-U.S. border. They both dedicated themselves to finding work as soon as they arrived. Dad has been

---

[1] A small town.

[2] A Mexican laborer allowed into the U.S. for a limited time as a seasonal agricultural worker.

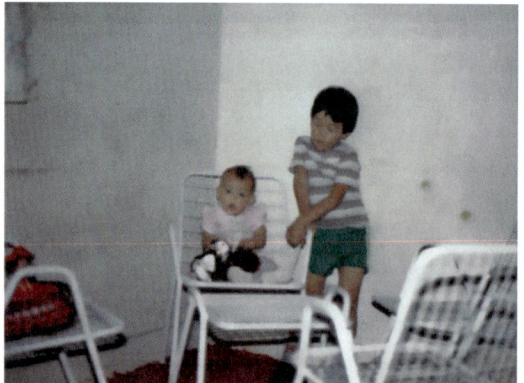 

Photos courtesy of Selenne Bañuelos.

(L) Picture of my older brother and me in 1985.
(R) Picture of my two brothers and me in 2016.

working in the same ground support equipment for air industries company for forty years. He began as a custodian, and thanks to his thirst for knowledge and perseverance, he moved on to material prep, welder, leadman, foreman, quality inspector, and production foreman. Mom worked full-time in the assembly line of electronics factories before my younger brother was born. Our parents left for work at about 4:00 AM every morning and would leave my older brother and I in the care of our neighbor. We lived in a duplex, so we shared a wall with a wonderful family who is still very dear to us. Our parents moved my older brother and me to the living room every morning when they left for work. Mom would leave a large boot with my older brother so that he could throw it against the wall if we needed to call our neighbor before 7:00 AM.

We grew up poor in a 900-square-foot home, but had other wonderful privileges like a large yard, friends with whom we grew up, a public library at the end of our street, books at home, and a support system with our extended family. Our parents emphasized the importance of education as a pathway to success. They bought us a children's encyclopedia that included a book on "World and Space." At nine years old I read about the Big Bang Theory and I remember wondering if the explosion I felt in my head while reading this was similar to the Big Bang. That explosion was the start of a whole new universe of learning about science for me.

## Early Years

I grew up in the busy Los Angeles neighborhood of Boyle Heights filled with vibrant colors, Spanish music coming out of stores, and the smell of taco stands. These colors, sounds, and smells could not mask the lack of service our community faced. According to a *Los Angeles Times* Boyle Heights profile based on the 2000 census, only 5% of residents 25 and older received a four-year degree.[3] Boyle Heights is also a redlined district that was then beleaguered in order to construct all the freeways that go through it.[4] I grew up in a home that is in front of the 60, 5, 10, and 101 freeway interchange. Our public school

[3] maps.latimes.com/neighborhoods/neighborhood/boyle-heights/
[4] tinyurl.com/rdong9u; https://tinyurl.com/y5v58z3z

 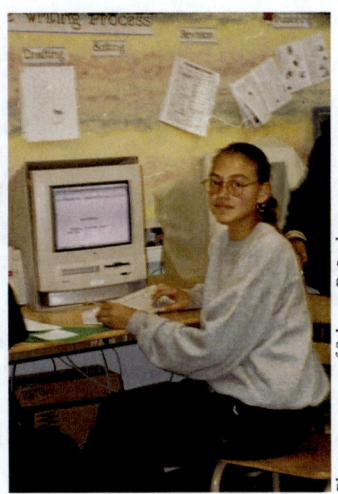

Photos courtesy of Selenne Bañuelos.

(L) A picture of me in fourth grade in 1994. (R) A picture of me in middle school in 1999.

system was also poorly served. The public high school I was assigned to, Roosevelt High School, was the second most populated school in the country at the time. Gang violence was also a major problem for those of us who grew up there. Navigating this life was tough.

Thankfully when our younger brother was born, Mom began working as a self-employed sales representative which allowed her to be present in our lives *y nos tenía con la cuerda bien cortita.*[5] Along with making sure we were safe, Mom had to learn to advocate for us as students. She tells our family that I had always loved school. As a first grader, I was always the first to get home and do my homework and the first to wake everyone up in the morning to go to school. So, when I started crying every morning saying I didn't want to go to school, she knew something was wrong. She decided to drop me off, walked around the building, and spied on our class through the window! She saw that a small group of students and I were being placed in the back of the room and ignored for most of the class time. Surely enough, she saw me crying in class. Mom walked into the office and gave the principal an earful. The rest of my time at this school was much better. I excelled in my classes and was asked if I wanted to skip third grade. My family and I decided not to.

**Middle school**. During my fifth grade, we lived in Mexico for one year so that my siblings and I could attend school there and improve our Spanish. Upon my return, I was ready to enroll in the local public middle school for sixth grade. I spoke with the school counselor who wanted to place me into nothing but development courses. Little did this counselor know that in fifth grade in Mexico, we were doing what seventh graders do here. I channeled my inner Chicana fighter—the same that I felt when watching *Stand and Deliver* as a young child—and told this counselor that he would be wasting my time. I told him that I would take those classes for one semester and that I would visit him again at the end of the semester and he would need to place me in the correct courses. My courses were changed after the first semester, but I was never told about Magnet or Honors programs while I attended this middle school. The lack of academic expectations from us as students was palpable. I felt that all this school cared about was making sure we survived.

---

[5] It translates to "and she had us on a short leash," meaning that she kept a very close eye on us.

I recall being very upset during a so-called pep talk by a visitor to one of my classes. She told us about all the statistics facing our community, like how many of us would die because of gang violence and how many of us girls would be pregnant before completing high school. She then asked who wanted to go to college and I challenged her by saying that she was asking the wrong question. Why not ask us what college we wanted to go to? Even at the young age of fourteen I distinctly remember saying to myself, *this is what oppression is.* The sheer lack of expectations that we, Chicanx students from Boyle Heights and East L.A., would amount to anything.

Fortunately, I had the pleasure of meeting Mr. Mitchell. He was my eighth-grade fall-semester math teacher. Mr. Mitchell helped me solidify my love for mathematics and encouraged me to apply to a boarding school for high school. He told me that I needed to survive East L.A. and hoped that attending a boarding school would pave a path towards college. I applied to a Southern California boarding school and was invited for an interview. My mother, my best friend, and I traveled in my mom's shabby 1981 Honda Civic hatchback, which we called Paco, to this interview. My friend and I laughed on our way towards the admissions building as we looked back at my mother's car parked among shiny BMWs, Lincolns, and Jaguars. As we laughed, my mother told us that we were hurting Paco's feelings and that he wouldn't be happy. It was a funny moment, but it helped me quickly realize that I did not belong there. After a full day of interviews and tours, we were finally headed home. It turned out that Paco's feelings were definitely hurt. The vengeful car did not want to turn on! To put the cherry on top of what felt like an embarrassing day, my friend and I had to push the car downhill so that my mom could kick start it. We got stuck at the bottom of the hill and a campus worker helped us start the car again.

**High school**. I attended an all-girls Catholic high school where the graduating classes were between 80 and 100 students. This school did not have many resources, but it was the time in which I felt the most empowered. I persisted through family challenges at home. I was trying to study in a 900-square-foot home with seven people in it. For most of the time while I was growing up, we had at least two adult family members from Mexico staying in our home while they worked. Some slept on the floor in the living room, and the light would keep them from sleeping. I resorted to studying in the bathroom. More than ever, I had aspirations to go to college, and I immersed myself in my studies. I was involved in student government, volunteered at the Los Angeles county medical hospital, ran track and cross-country, and was a letterwoman. All first-years took Algebra I. Towards the end of the spring semester we were informed about the school goals to offer the first AP Calculus course. Those of us who wanted the opportunity would need to take Geometry in the summer, so that we could move on to Algebra II our sophomore year, Trig/Pre Calc our junior year, with the finale of AP Calculus our senior year. About thirty young women signed up for this challenge. Some of my most memorable moments in high school are studying for Calculus with my friends at my house while we used my parents' bedroom mirror as a whiteboard. These instances helped me set my goal to become a math teacher.

I wanted to empower students from my community through mathematics. Unfortunately, navigating the college application process was a very distant notion as no one in my family had attended a four-year college. The first time I stepped foot on a college campus was through an outreach program held at Loyola Marymount University.

 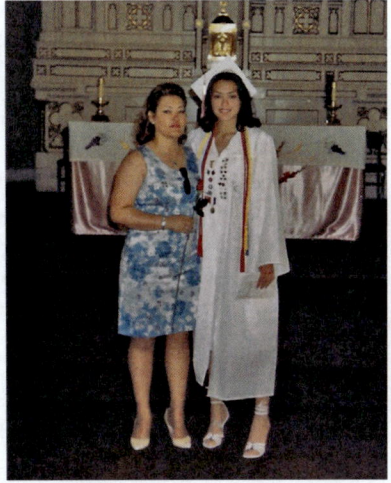

Photos courtesy of Selenne Bañuelos.

(L) Picture of me in my cross country uniform 2003.
(R) Picture of my mother and me for our senior year mass in 2003.

This was a two-week summer program held on campus for high school juniors from underrepresented minority populations chosen for their strong interest and aptitude for math and science. This program helped me feel confident in myself as a student. I applied for college my senior year and accepted the admission to the University of California, Santa Barbara (UCSB).

## Higher Education

**Undergraduate education**. I loved my time at UCSB. Sure, it was quite a big change from my high school—the small class sizes I enjoyed in my high school greatly contrasted the 300+ Calculus II class I took my first quarter. The class sizes, however, were not the only problems I faced. Not only was I one of the few Chicana students in my STEM courses, I constantly felt underestimated and unnoticed. After my first quarter at UCSB, I was speaking with a male student who asked me what my grades were for that quarter. He was very surprised that I earned a higher GPA than he did or that I did better in Calc II than he did. I said to him, "Of course I did; I'm a woman." I did not know that I was not *supposed to be* good at math because I am a woman or because I am Mexican-American for that matter. This bias was not touted in my high school. We were all women, the majority of which were Latina, and we did great. Later that same school year, after earning the highest score in a midterm for our Ordinary Differential Equations course, a student from the creative studies college claimed that he would have earned the higher score if he cared about his grade. There are a lot of "little" stories like this sprinkled throughout my time as an undergraduate in mathematics. They did start feeling heavier as time went on, but once I formed a strong connection with other math students, I learned to move past those microaggressions.

I went to UCSB with the intent of getting a mathematics degree in order to go back to teach in my neighborhood. However, during my junior year, a professor, Dr. Kenneth Millett, saw potential in me and many times suggested that I continue my studies. I would tell him that there was no way I was going to get a PhD; my plans were to be an influential math teacher. We would actually argue about this quite a bit. He was worried that I was not letting myself consider a doctoral degree as a possibility, but I was truly concerned about not having a place in *that* world. At the time, I was having an identity crisis. My family life—the musical, vibrant, tequila-drinking side of me—seemed to clash with what I perceived a mathematician's world to be. I also looked around the math department and did not see a tenure/tenure-track woman nor an American-born Latinx. I wondered: what does it matter if I get this degree if there is no place for me here? Dr. Millett understood where I was coming from and said that I could help change that. Dr. Millett also said something that I say to my students now. "No matter what your plans are, there is no reason why you should shut doors to programs or opportunities that come your way; you never know what you will discover about yourself."

I followed Dr. Millett's advice by taking a graduate course, and with his and Dr. Jeffrey Stopple's help, I attended the Conference for Undergraduate Women in Mathematics at the University of Nebraska, Lincoln. Professors Stopple and Millett succeeded in sparking my interests. During the conference, I became curious about research in applied mathematics after listening to many women my age give presentations on the work they had done in summer research programs. I then decided to apply for an REU (Research Experience for Undergraduates) to attend that summer and was fortunately accepted to the Summer Math Institute (SMI) at Cornell. SMI was an intense eight-week program designed to help prepare participants for the rigors of graduate school by having a course in real analysis and a research project. I was consumed in mathematics from 8:00 AM until midnight every day and I loved it. This program also helped me realize that I had the endurance that a graduate program demands.

I'll never forget getting back to UCSB that fall and talking with Dr. Millett. He asked, "Well?" I said, "I'm thinking about applying to graduate school." He smirked and nodded his head. Till this day I can see him patting himself on the back. However, I still wondered if I belonged. That is where the Society for the Advancement of Chicanos and Native

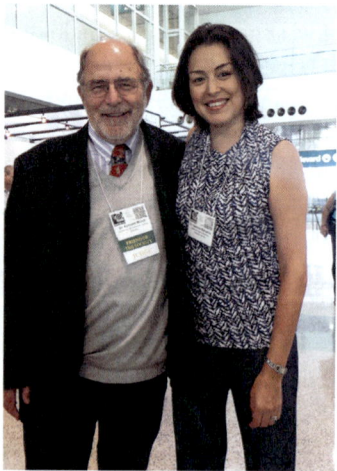

Photos courtesy of Selenne Bañuelos.

(L) Picture of the SMI group at Cornell in the summer of 2006. (R) Picture of Dr. Millett and me at the 2014 SACNAS conference.

Photos courtesy of Selenne Bañuelos.

(L) Pancake party at Dr. Stopple's house for graduating students in math (Dr. Stopple is the third person from the right). (R) Picture of my parents and me at my BS graduation at UCSB in 2007.

Americans in Science (SACNAS) organization made a lasting impact. The research mentors at SMI had applied for us to attend the SACNAS conference that October. In October of 2006, the fall of my senior year at UCSB, I attended my first annual SACNAS National Conference. It was a life-changing experience. I walked into a room of 3000 scientists, academics, and professionals who looked just like me and who had lived through similar struggles as my own. SACNAS gave me that push I needed. I attended the many professional development sessions that were aimed for students applying to graduate school. I learned of the application process and what things to look for in a program. Attending this conference helped me make the decision to apply to graduate programs. I knew that I had formed an academic family, one that would support me and help guide me as I took on this endeavor.

**Graduate education**. I was fortunate to gain acceptances to several graduate programs and decided on attending the University of Southern California in Los Angeles. My early years as a graduate student were the most difficult of my career. I felt that a mistake was made in admitting me and that the program was not prepared for a student like me. I was honored to receive a Provost Fellowship which funded my graduate studies and included a reduced graduate teaching load. However, it is the mentoring and the support network that I needed the most. In an incoming cohort of 20 students, there were only four of us that had been admitted with bachelor's degrees from an undergraduate institution in the United States. The rest of the students were those with master's degrees or international students. Most of the students were much more advanced than I was even though I had taken a graduate course and conducted research as an undergraduate. This made it very difficult to find a group of students to work with during my first year. I questioned my future in the program every day. Fortunately, I completed my coursework and qualifying exams. At the same time, I had formed a support network outside of the department with Latinx graduate students in STEM programs. We formed the first SACNAS chapter at USC. While working on my research, I also felt quite isolated. My advisor was the only tenure-track faculty member working in this research area, and I was his only graduate student. There were no seminars that I could create with peers so that we could help each other advance in our work.

I was a graduate TA from my third year through my sixth. I was happy to learn from this experience that I truly loved to teach. I was awarded the Department of Mathematics

Photos courtesy of Selenne Bañuelos.

(L) Picture at my PhD hooding ceremony in 2013. (R) Picture of my family at Yosemite in 2018.

Denis Ray Estes Graduate Teaching Prize and was the single departmental nominee for the University Outstanding TA Award.

During my fourth year of the program, I married my husband, Raul, and we welcomed our son a year later. We supported each other through our new role as parents, and he was incredibly encouraging as I continued my studies. I now had an even stronger motivation to complete my degree and pursue my career as an educator. Soon after graduation, we learned that we were expecting our second child. I felt that it was too much to move around and was supported by the USC mathematics department as I obtained a lecturer position for a year. I then traveled with my four-week old daughter across the country as I interviewed for positions, and I am very grateful for those institutions that adapted their schedules for me and my daughter.

**Some advice for students regarding graduate school**. I knew that graduate school was going to be very difficult, and I felt that having my family nearby would be important. However, it did not matter that my parents were just a drive away. All they could do for me when I was stressed and thinking about quitting (which was daily), was say "it'll be OK, *mija*."[6] They could have done that over the phone. I ask students to think about how important they think the location of the school is versus a program that is intentional about mentoring their graduate students, that cares about learning who their graduate students are, and creates opportunities for their professional development while they obtain their degree.

The research that is conducted during your PhD is not the be-all-end-all. I am one of several faculty who has changed their research focus from their dissertation. There are some overlaps, but the majority of the research that I have conducted over the past six years is not related to the work I conducted as a graduate student.

---

[6] My daughter.

# Current

**Research**. My research interests lie in the fields of differential and difference equations, dynamical systems and its applications to mathematical biology. I am interested in the insights that mathematics has on biology and vice-versa. I have had the privilege of working with biologists who express how they would like mathematics to inform their work.

I consider myself incredibly fortunate to have participated in the Women in Applied Mathematics research collaboration group that was hosted by the Institute for Mathematics and its Applications (IMA) at the University of Minnesota the summer I completed my PhD. I joined a group of six female mathematical biologists and mathematical modelers to conduct research on the effects of thermoregulation on sleep. Human sleep is divided into two physiologically different stages: rapid eye movement (REM) sleep and non-rapid eye movement (NREM) sleep. NREM sleep is a state of reduced brain activity which includes deep slow-wave sleep. REM is the stage where we have our most vivid dreams and our brain metabolism is as high as during wake. However, REM is distinguished from both NREM and wake by an absence of muscle tone and a suspension of thermoregulation. In our work we develop a model for REM and NREM dynamics with sleep/wake cycling by building on recent neurophysiological models and answer questions regarding ambient temperature and human sleep. I am especially grateful for the friendships that have been formed with Drs. Shelby Wilson, Alicia Prieto-Langarica, Pamela Pyzza, Gemma Huguet, and Janet Best. The support and camaraderie in this group has been essential for my success.

Another project I am involved with is in the area of mathematical epidemiology. Zika is a vector-borne disease similar to Dengue or Chikungunya; however, due to the possibility of sexual transmission, other models are no longer applicable. We created a model of the spread of Zika that was then used to determine the basic reproductive number with and without the effect of sexual transmission; we estimated unknown parameters via Latin Hypercube Sampling and analyzed the sensitivity of the parameters. We also implemented a simple version of control using Wolbachia bacterium the analysis of which was conducted by the targeted reproductive number. This work was supported by two SACNAS

Photo courtesy of Selenne Bañuelos.

From left to right: Pamela Pyzza, Alicia Prieto-Langarica, Janet Best, Shelby Wilson, Gemma Huguet, and me at the Institute for Mathematics and its Applications in 2013.

collaboration mini-grants and I also guided two master's students through their thesis in this work.

I advised a group of seven undergraduate students in a research project along with Dr. Cynthia Flores during the 2016–2017 academic year. The research project explored the dynamics of classes of voters by taking a mathematical epidemiology approach. Guiding this group of undergraduate students really helped me in my fundamental understanding of modeling the spread of diseases, which aided our Zika virus research. I learned where to begin guiding future research students on epidemiology projects.

## Final Thoughts

For some time I believed that all first generation, Latinx, or low-income students needed were opportunities and *ganas*[7]. But, after years of reflecting on my upbringing and my career path, I noticed that many things had to align for me to get to where I am. Growing up in Boyle Heights I could have been in the wrong place at the wrong time as were some of my friends. I also ask myself, what would have happened if Dr. Millett or Dr. Stopple had not taken an interest in my future? I could have stopped at my BS degree without knowing that I was capable of more. How many students don't have a Dr. Millett or a Dr. Stopple? I opine that STEM departments in higher education should have systems in place for students to be mentored so that this work does not fall on only a few faculty.

I am often asked by students how I shook off the impostor syndrome. My reply is always the same: I have not. I asked a mentor of mine who I greatly admire the same question. I was very surprised to hear that she still feels the same way. I was flabbergasted. Here was this amazing leader, incredibly hard-working, equity-minded mentor saying she hasn't shaken it. I had a range of emotions because of her honest response. I first felt sad thinking that I will always feel like I do not belong or that my work is never enough. But after a lengthy conversation with her, I learned how to channel that emotion and feel empowered. Maybe the impostor syndrome doesn't stop, but I have not let it stop me either.

Some of you may also receive sly remarks insinuating that you have not earned your position due to your ethnicity or gender. Here are two comments I have for that: 1) I have been fortunate in having choices for positions along my career. I knew I would earn a job somewhere and there will always be at least one person in each of those places that will have that thought. 2) If we operate under the assumption that I was hired because I am a Mexican-American woman in mathematics, it is not the reason why I have earned the retention each year. My teaching, research, service, and viewing mentoring of underrepresented students as an essential part of my career are the reasons.

---

[7] The will to succeed.

# 4

# Dr. Erika Tatiana Camacho

## From Mexico to Los Angeles

My dad died when I was three months old leaving my mother to care for me and my three older siblings, ages four, three, and two. This was in Guadalajara, Mexico, where a lack of money meant lack of opportunity. After struggling to raise us by herself for a few years, my mother took up the offer of my grandma (my dad's mom) to take the four of us in, so that she could focus on working to make ends meet. My grandma lived two hours outside the city and so this also meant that my mom was only able to see us on the weekends. Worse was the fact that my grandma was not a caring person. Behind closed doors, she allowed us to be abused often and it seemed like she just considered us free labor. My mom would give her most of her paychecks thinking that the money was going to our clothes and food but this was not the case.

Dr. Erika Tatiana Camacho

Illustration created by Ana Valle.

We didn't have any toys to play with and many times would go hungry. She would also have us sell candy on the street and other things that my mom was not aware of. After a few years, my mom finally caught on and took us back to Guadalajara with her.

When I was seven, she met my future stepdad and corresponded with him for a while. He lived in the U.S. and promised an opportunity for a better life. After nearly a year of correspondence dating, he proposed and we soon moved to the U.S. Unfortunately, he had not been honest about his financial situation and we (my stepdad, mom, me, and my three siblings) first moved in with his adult nephew for a year and then into a one-bedroom apartment in South Central Los Angeles. Life was tough as we didn't speak English and went straight to English-only school. There are many stories of the rough life we endured for the next year in school, walking to and from school, and around the neighborhood. For example, my brother got stabbed within the first week of getting to the U.S. by two 18-year-old guys who followed us when we were walking from school (all I remember is him telling me and my sisters to run as he thought he would hold them back). After two

years, we moved to East Los Angeles with my
new baby sister to an equally harsh environment but at least we knew the language since
most people spoke Spanish. We remained in
East L.A. (in the same two-bedroom apartment) until I was already in college.

**Shaping my dreams—the teachers who
inspired me**. It was my time in East L.A.
that really formed my academics. I could not
speak English that well, in part because the
language that was still spoken in my house
and community was Spanish. In seventh
grade, my overcrowded school mistaken-
ly placed me in Honors English instead of
English as a Second Language (ESL) class.
When I tried to talk to the school counsel-
or, I was told that I should be happy that I
was placed anywhere because at least 100
students still did not have a placement. The
Honors English teacher, Mrs. Warnock, who
noticed that I was not going to get placed in
ESL after multiple times of trying, told me
that if I was willing to put in the work during
lunch and after school that she would spend
the extra hours tutoring me. I accepted, did
well, and was placed in Honors English from
there on (thus, by chance, ensuring that I
would have the required classes to apply to a
four-year college).

In high school, I was fortunate to have
the late Jaime Escalante of *Stand and Deliver*
fame as my math teacher. He was the one
that really showed me the power of math and
what it can do with hard work. While I was
not featured in the movie (it was based on a
class ten years before me), the depiction of a
tough-love teacher who had Saturday morning math sessions was very precise. Before
meeting Jaime, my dream job was to be a cashier. Jaime, or "Kimo" as we called him (for
*Kimo-sabe*, the one who knows it all), often
brought alumni into the classroom. One of
these alumni was a student who went to MIT
and worked at a research lab in California.

My *mami* and I celebrating my oldest sister's
birthday in 2020.

Giving a speech at Jaime Escalante's (Kimo's)
memorial.

He talked about what engineers can do with math and about the nice car that he drove and the peaceful neighborhood he lived in. What was an old dream changed: I wanted to be an engineer and go to MIT!

## College Years and Beyond

**Struggling at Wellesley.** My high school student government sponsor, Mrs. Dumas, gave me $500 to apply to colleges when she found out from other students I didn't have the money and thus was not planning to apply to any college that had an application fee. Being too proud, I told her I couldn't accept it and I only accepted it after she insisted and put the condition that I would not repay it to her but would instead "pay it forward," which is something I feel I have spent much of my life after college doing with numerous students and individuals with similar struggles. I went to Wellesley College for undergraduate school and wanted to be an engineer and take advantage of the joint Wellesley-MIT dual engineering degree program.

In my first physics class at Wellesley, I struggled, partly because of the language, but mainly because of the racism of both the professor and my lab partner. When I went to office hours to ask for help on a particular topic, he started off by asking where I was from, and then after I answered, proceeded to say that he did not understand why they kept letting people like me in while taking a spot from other more qualified applicants.

During one of my first physics labs, my partner asked if my accent was French because she had been to France a few times. When I told her my accent was because I was Mexican and I came from East L.A., she said that she could not be my partner because I would hurt her grade and she immediately went to talk with the professor. She was promptly reassigned to a new group—everyone was in pairs except her group of three and my group of one. When I asked the professor if I was going to have another lab partner assigned

Celebrating in college with my friends from high school Becky Marquez and Christina Miranda.

because it was often hard to do the experiments alone, he repeated the same comment he told me in his office hours and said that he could not blame my lab partner for not wanting to work with me. We had not had a single thing graded up to this point, so I internalized this and began to question if I really belonged there. Either by myself or occasionally with a Spanish major friend helping in the late nights when no one else was in the lab, I struggled through the semester. I was devastated by my C grade and realized that my dreams of being an engineer were gone.

I was always very good at and enjoyed math so I decided to be a math major. I had also taken an economics course and really liked it so I double majored with economics too (and econ was one of the few classes in the sciences where the students were always willing to work with me). Even though I was good at both subjects, I still had to work really hard. There were many times when I was ready to quit. One of the hardest parts was not feeling that I could call my mom because her show of support would be to tell me to come back home, leave college, and that one of my sisters could help me get a job.

When I was in this situation, I would sometimes call Kimo. One of the times I called him, I reminded him that he promised me he would get me to MIT but that I was at Wellesley, and things were rough. He livened the moment immediately by telling me that his estimate is not an exact science. He said he did really good by getting me within 50 miles of MIT since Wellesley is just outside of Boston. He also reminded me of why I needed to stay in school. Key mentors, like Kimo, that supported me in critical moments were what helped get me through undergraduate school.

**Towards a PhD in mathematics.** One of the things that inspired me to pursue my PhD was being able to participate in a summer Research Experience for Undergraduates (REU). It was after this that I decided I wanted to become a professor, and impact students through research and mentoring.

In grad school, I remember studying analysis with a Latino friend and we were thrilled to be taking the class from the only Latino math professor. We did our homework together and I consistently did better than him. When it came to the exam, he scored higher than I did by a letter grade. On the second exam, I again scored a letter grade lower. I went to the professor's office to go over a few mistakes I made on the exam. The professor took this opportunity to belittle me and to lecture me on how it was important to do my own work and not to cheat. I asked him what he meant and he said that I was clearly copying from my friend's problem sets since we had about the same homework score but that he scored higher than me on the exams. I was devastated by the accusation. It was hard to sit through the class for the rest of the semester and the following one too (as I had the same professor). What was most disheartening was that my friend in the class confided to me that he had been given the exams from the previous year and that the questions were almost identical. He did not offer this material and I did not ask him to share it with me nor did I ever tell the professor. I just started to distance myself. A year later he asked why I did not want to work or study with him anymore. I told him about what the professor had told me and he laughed and said it was funny that I was accused of this.

During my third year in graduate school I had my first child. My husband was three years ahead of me, and had just graduated. We tried to move back to Los Angeles as he had a job lined up and I thought I would be able to do my research remotely. After a few months of a lack of productivity for numerous reasons, I realized that if I was going to complete my PhD I would need to move back to Ithaca to do it. It was one of the hardest decisions I had to make. However, seared in my mind was my personal upbringing of my mother, who only had a very limited education, working two to three jobs to support her family because you never know what life will throw at you. I wanted to make sure my son was never in that same position and so I moved back to Ithaca and he stayed in L.A. with his dad (and my mom and family). It was a very difficult two years to finish, but I saw no other way to do it. I was so proud to have earned my PhD, but it came at such a high price.

**Early experiences as a professor**. In my first permanent job, I was struggling with the lack of respect from students in the service courses I taught. I had a student that, upon receiving a well-deserved C on a test, stood up in the middle of class and proclaimed that his tuition paid my salary and that I needed to curve the exam because he needed a B in the course. I told him that he deserved the grade he got but that he could come to office hours and I would help him. He used profanity and told me I would be sorry I gave him a C and that he would go talk to my Chair, which he did. I was then called to my Chair's office and lectured that I couldn't tell the students that they deserved a certain grade and that I needed to schedule more time to work with this student than just office hours because those hours did not work for him. Grudgingly, I obliged, but the student rarely kept appointments we scheduled. From then on, he would use profanity when referring to me in the classroom, which my Chair refused to address but instead reminded me that I could not kick the student out or do anything but continue to be polite and teach.

Not much changed my second year, except for my Chair. One day, she called me into her office and said that I needed to stop wearing skirts and bright colored clothes and this was the root of all behavior problems in my classes. She explicitly told me to start wearing suit coats and pants and to make sure they didn't fit tight and nothing with bright colors. Her basis was one comment in one student evaluation that said he couldn't concentrate because I was too cute and young to be a professor. Yet she ignored the multiple comments from students who explicitly wrote "send her back to Mexico," "get a professor that knows how to speak English," "we cannot understand her thick accent," etc. and from others to kick the disrespectful kids out of class. I was devastated but found no other choice than to listen.

There have been many times that I have felt like quitting and multiple times, I was one conversation away from just walking away. In looking back, the striking thing is that it was never about mathematics! It was about the climate and culture at a place, stereotypes that people held, subtle and explicit racism, and microaggressions I endured and to which I am still subjected to this day on a regular basis. I think many people believe that academia is a blind place where your talents will be recognized appropriately and where you will not be judged by your looks or other characteristics. But the reality is that academia is made up of people and those people have their biases just like they do everywhere and is one of the most judgmental, inequitable, and political places. Too many people use "merit" to cloak discrimination. While I have had some terrific mentors, both male and female, who were Caucasian, it was hard to go through this process rarely being taught by or working with someone who looked like me.

# Research—Applying Mathematics to Understand Vision

I study mathematical physiology, specifically looking at components in the visual system that lead to blindness. I collaborate with experimentalists and together we try to understand what causes blindness in diseases where the photoreceptors degenerate, such as in Retinitis Pigmentosa (RP). I have focused much of my attention on the photoreceptors in the retina, the rods and cones, together with the retinal pigment epithelium (RPE) that works with rods and cones to facilitate vision.

I am the first person to dive into this area from a mathematical perspective. My research and publications on the subject provided the first set of mechanistic models and mathematical equations describing photoreceptor degeneration. I have 28+ publications with 12+ of them pioneering modeling of retinal processes. We developed a series of spatially averaged nonlinear ordinary differential equations models to investigate both the healthy and diseased retinas at the cellular and molecular levels. In my earliest publication in the area, we predicted the existence of something experimentally discovered a year later—the rod-derived cone viability factor (RdCVF) and proposed equations describing the dynamics of rod and cone outer segment (OS) number, and RPE cell number. We extended this model in another publication to account for mutated (phenotype) rods, which were used to describe the diseased state of RP. In another publication, the RP model was further extended to include a control input that represented RdCVF treatment, which was designed using optimal control. The metabolic contribution in RP to photoreceptor death was quantified in collaboration with an experimentalist through a mathematical model. With another experimentalist (the discoverer of RdCVF, Dr. Thierry Léveillard), we showed the role of RdCVF in RP coexistence and mathematically analyzed it.

My recent work extends to multi-scale predictive models that investigate cellular and molecular level dynamics contributing to degenerative process or diseases including Age-related Macular Degeneration (AMD). I want my research to make a difference. Being sought out by some of the top experimentalists in the world for collaboration on how to stop photoreceptor degeneration (and mitigate blindness) has been a huge recognition from my point of view. It was incredibly rewarding when my collaborator Dr. Léveillard said it was because of our models and my discussion with him about their predictions that his team was able to identify previously unknown metabolic pathways that are now under investigation.

# Mentoring and Perseverance

I see so many things in academia and outside of it that need to change to make an equal playing field, yet it seems that most people in positions of power are not willing to risk it to make a change. I feel like there are "fights" at nearly every step that would lead in the right direction.

I am currently in my reflection stage and not sure how to best change academia and the world while not giving up all of me, especially given that there is no guarantee that what I do might make any noticeable difference. It's tough. What often keeps me going are the many, very personal, and sincere messages I receive from former mentees of how I made a difference in their trajectory. The most touching ones are when they have told me that

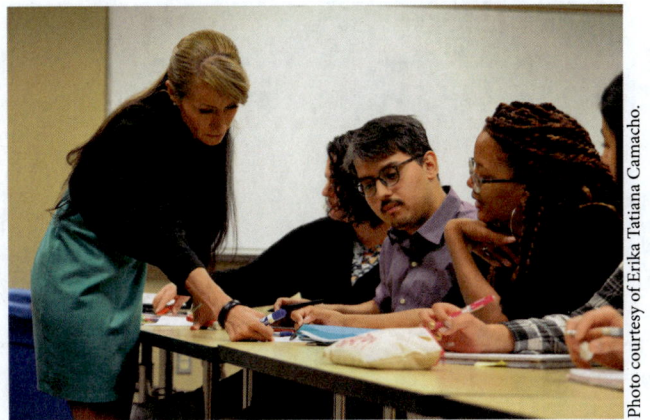

Mentoring Roberto Alvarez, Danielle Brager, and students at ASU in 2019.

only after being away for a few years did they really appreciate the mentoring I gave them. Many will ask what they can do to become a mentor like me because they have not found people like me. I explain that I just do what I learned from my mentors who really truly cared about me—those who did not mentor me because they were trying to build an empire or check a box. I have fought many battles for my mentees when they were experiencing bias or unfair treatment by jumping in and changing the situation for them before they were completely broken. I have also fought for many who are not my mentees but who don't have a voice or seat at the decision table or who are invisible and forgotten, often by making structural changes and creating opportunities. I sacrificed many things to give my mentees and many others equal access and opportunites. I try to do whatever I can to help my mentees realize their full potential and their dreams, in addition to listening and giving advice that I think is best for them irrespective of me. Each one of their successes gives me the motivation to continue. Each one of the mentoring awards I have won is also a validation that what I do actually makes a difference.

Receiving the American Association for the Advancement of Science Mentor Award in 2019.

Receiving the Presidential Award for Excellence in Science, Mathematics and Engineering Mentoring (PAESMEM) award in 2018 with Frances Cordova Director of the NSF and the White House representative of Science Policy.

A lot of mentoring is first learning about the individual because so many factors influence who we are today and why we make certain choices. Then it is a time-intensive endeavor to meet them where they are and to bring them up to their full potential.

As a postdoc, I co-wrote a grant to the National Security Agency (NSA) and the National Science Foundation (NSF) to start my own REU. This led to the launch of the Applied Mathematical Sciences Summer Institute (AMSSI) the summer after my first year in my tenure-track job at Loyola Marymount Univeristy (LMU). The program was joint with Cal Poly Pomona and we split the time between the two universities, moving the students halfway through the program. Over its three years, we had 48 students, and five Teaching Assistants (TAs) participate in the program. All were either from underrepresented backgrounds or schools where research opportunities didn't exist or were not available to them. Nearly one-third earned their PhDs, with many others earning their master's. It was a very rewarding program and only stopped because I moved from LMU to Arizona State University (ASU).

I have given motivational talks to groups ranging from elementary school through high school students in addition to college students, professors, and high-level administrators. Each of these talks can get emotional as I'm giving examples from my life to drive home the message. One of the most emotional keynote addresses I gave was in 2009 to the recipients of the Presidential Awards for Excellence in Science Mathematics and Engineering Mentoring (PAESMEM). It was very special to be able to thank each of them on behalf of all their mentees and tell them how their efforts are appreciated even if not all the mentees come back and say thank you. It's also important to keep in mind that great mentors are not just the ones that have won the big awards, but most important the ones who open the doors for others by creating opportunities and fighting the tough battles to bring about change.

## Diversity and Inclusion in Mathematics

I am an applied mathematician and thus I usually see everything as a problem to be solved. I really have come to appreciate what is now known as "team science" where multiple people come together from their varied perspectives to try to solve a previously unsolved problem. I think the most challenging problems require this approach.

We need a society that is knowledgeable about STEM or at least appreciative of it! Since math describes and underlies nearly every aspect of STEM, every issue can be helped with the presence of mathematicians. Moreover, we really need to focus on team science where multiple approaches, even beyond STEM, are used to solve the most pressing problems of today.

In higher education, we talk about changing things for marginalized communities, yet we forget to include stakeholders from these communities (we bring high-level administrators or researchers to the table as "representatives" who are employed in institutions serving these communities but who have not been raised in these communities or have not lived through the experiences of those they represent).

The mathematics students and faculty need to reflect the U.S. population. I think one of the main barriers is a lack of understanding from the professors about what it means to be a professor. We judge people by their publications and not the impact of all their efforts (scholarly and otherwise). If we want to change academia and agree that we need all professors to be good mentors and to understand the diverse population they are working with, why should people object to requiring a Diversity/Inclusivity Statement and a Mentoring Statement in job applications and promotion documents?

From a different angle, individuals rarely solve the challenging problems of today. If we let those perpetuating the status quo also determine the research agenda, very little will change. We need to bring the richness of multiple backgrounds, including Latinx people, to be at the table and set the research agenda, yet it's understandably hard for many of those in charge to step aside and let different perspectives weigh in.

In terms of societal/institutional structures, it's the narrow-mindedness, micro-aggressions, and institutional racism that are the biggest obstacles to learning.

It will take many selfless and courageous people of every race, color, and way of life to eliminate these structures because there are too many vocal people (even if they are in the minority) that want to preserve the status quo or go back to how things used to be and there are too many people who don't think it's their problem and will stay silent. We have a generation of Latinx PhD mathematicians that wasn't present when I started my journey and that really gives me hope that change is on its way. Some are oblivious to the Latinx situation but most are actively doing things to promote our community. Many majority mathematicians show us support too.

We need more Latinxs involved in math because of the need to approach challenging problems from different perspectives and to actually shift the focus of what problems we should be working on.

## Advice

It's great to want to change the world. But don't lose yourself as you try to do it. There will always be detractors and haters, but ignore them as much as possible. Don't doubt yourself and don't let others define who you are or your path. Judge people by their actions and not just their words. Finally, while it's great to have a mentor that looks like you, don't think that this is a requirement. Some of your very best and long-term mentors may be Caucasian males, and one that may do the most damage to you personally and professionally might be a Latino.

Many times I am speaking to students who have had a rough path yet they have what it takes to overcome the adversity they have experienced. I had, and still have, a rough journey many times. Sometimes they just need to realize that people have done it before, are still struggling to make it in their respective career levels, and they are not alone. The path won't often be easy, but the rewards, in the end, are worth it.

I have had some terrific and some very hurtful and toxic mentors who are mathematicians. But I have also had great mentors that are not in math and some are not even in STEM, but they are very perceptive, understand things, and can give relevant advice. It's them needing to realize who I am as a person and what may be best for me. Many people from all walks of life will share and support your goals and dreams. Perhaps there is a correlation between those supporters and people with your characteristics, but by no means should you limit your mentors and advocates to just those with certain characteristics (such as being Latinx). At the same time, believe someone the first time they show you their true colors.

# 5

# Dr. Anastasia Chavez

## Early Life and Finding Mathematics

There are three jobs I ever remember wanting as a kid: professional clarinetist, U.S. Olympic softball player, and first woman president. Talk about #careergoals! As you undoubtedly noticed, being a mathematician did not make this ambitious list. In fact, a career in math was not an option I was even aware of until college. Before then, it was just a subject I enjoyed (mostly), was good at (mostly), and kept pursuing even when I thought I wanted to be a writer and journalist.

In hindsight, there are hints of a love for mathematics in my childhood: I enjoyed puzzles enough to have a dedicated piece of plywood that I kept under the couch; I loved to do word searches, unravel riddles, and read problem-solving mystery books; and being fast and correct on timed math

Dr. Anastasia Chavez

Illustration created by Ana Valle.

problems was one of my favorite pastimes in grade school. Though, if my family history was any indication of math pursuits, one would have easily written my potential off as a nonstarter. My father's parents were Mexican immigrants with a grade school education, I had no family members with a degree beyond high school, and I presented a dark complexion, none of which historically are in a U.S. mathematician's favor. Yes, my father worked in construction and inevitably had to deal with geometric deductions, and my mother's family has a lineage of talented seamstresses and dental technicians. However, as I was pursuing my doctorate in mathematics at UC Berkeley, a common phrase my mother would say was "I don't know where she gets it." Clearly, math just didn't run in the family.

I grew up in the town of Santa Rosa, California, the eldest of three daughters. From a young age I loved sports, particularly baseball, and I loved proving my ability against anyone willing to test me. This competitiveness kept me driven to hit further, throw harder, and run faster, especially against the boys. It even got me into a few near-fist fights on the playground as I flexed my skills to prove no boy could beat me at wall ball. In hindsight, my determination to succeed in comparison to men was instilled by my dad. He would often tell my sisters and I that being a brown woman was an asset, not a fault. He taught us how to use power tools, build decks, paint houses, and showed us our worth by hiring us to work alongside him on construction sites. We could see his pride in our skills when, while chuckling, he told us the other men on the job complained we were too fast and made them look bad. I learned to value myself as a woman of color and recognized that not everyone did, Latinxs and non-Latinxs alike.

**Family, culture and upbringing.** Latinx immigrants are in my blood: my mother's great-grandparents left Spain for the U.S. during the early 1900s as dictatorship was on the rise and World War I was looming; my father, a first-generation Mexican-American, grew up the son of poor undocumented apple pickers. Yet, I was raised speaking only English and with little exposure to Latin American culture. During the late 1980s and 1990s, this kind of conscious assimilation was common, encouraged, and even mandated (see California's Prop 227 of 1998). Although I didn't identify with Latin culture or its language, others inherently assumed it for me. This kept me an outsider, too brown to be American and too white to be Latina. During a seventh-grade history class, I saw the disappointing judgment of my white teacher when, after she asked me to translate a homework assignment to my Spanish-only speaking classmate, I admitted I did not

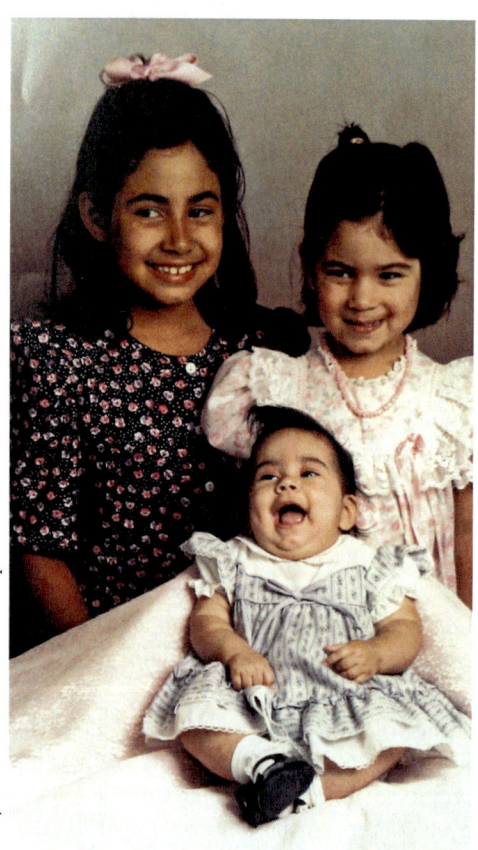

Photo by Sears Portrait Studios/CPI Corp.

The Chavez sisters.

speak Spanish. It stung to hear a group of Latinx high school classmates laugh when they heard my American accent coming through as I spoke Spanish during our language class. In spite of this culture-shaming, I have found acceptance of my hybrid Mexican-American experience while strengthening my identity as the version of Latina that I am. In fact, mathematics helped me realize the importance of cherishing my own heritage and the culture of my ancestors. Sadly, it has been through the lack of Latinx representation in mathematics, and specifically U.S. born Latinxs, that spurred this awakening.

Where I did see Latinx people was in literature. As a teenager, I was mystified by Gabriel García Márquez, Federico García Lorca, and Sandra Cisneros. The colorful fantastic worlds they developed revealed a creativity and imagination of which I wanted to be a part. I discovered while volunteering at the local downtown library that the quiet vastness of literature and the buildings that housed them were one of the most ideal places to be. I often envisioned my perfect life as living in the library, devouring book after book, sustained by their knowledge. So, after haphazardly enrolling in the Santa Rosa Junior College after high school, I pursued my two favorite subjects: writing and sports, with mathematics as an afterthought. It was my calculus teacher, Kirby Bunas, who was the first person I can remember to tell me directly that I was good at math. She even hired me as a grader for her class the following semester. This was huge for my mathematical confidence, but did not stop me from showing up late for a calculus final in my pajamas three semesters later. I still considered my argumentative and creative writing courses top priority. In fact, I might never have pursued mathematics had it not been for another haphazard choice. Upon deciding to transfer to San Francisco State University (SFSU) to major in journalism, it was a guidance counselor who pointed out the math grades on my transcript as evidence of my preparation to major in math. He then said, "You can do anything with a math degree." To which I said with a shrug, "O.K."

Me in my baseball uniform, age 9.

## Entering the World of Mathematics

Only now can I see that my undergraduate years exposed me to all the "must haves" of a student whose path led to a doctoral program: supportive mentors, research experience, mathematical breadth, and community. I met one of my first true mentors, Dr. David Ellis, a no-nonsense African-American applied mathematician, who helped me apply for and receive enough funding to pay for my education the remainder of my time at SFSU. Thanks to Dr. Ellis' encouragement, by my second semester I was accepted into the Undergraduate Biology Mathematics Collaboration (UBM) program. For two years I was funded to do multidisciplinary research in mathematical biology, co-advised by Dr. Ellis and Dr. Edward Connor, a professor of ecology. I developed coding skills using software such as *MatLab* and *Mathematica*, took courses in biological statistics, and collaborated with both undergraduate and graduate students.

I gave my first research presentation on this project at a campus-wide poster competition and to my surprise received fourth place. This research experience allowed me to develop a mathematical identity and confidence I never had before. I felt supported and valued by my professors and peers, and was encouraged to pursue graduate school. Again thanks to Dr. Ellis, while I prepared graduate school applications, I also applied for the CSU-LSAMP Bridge to the doctorate program through SFSU's master's program. In writing my personal statement, I realized my undergraduate experience not only exposed me to critical experiences to develop as a mathematician—it also exposed the need for Latinx and women in mathematics. My first Latinx mathematics professor wasn't until my final semester of undergrad, and I've had only one other since then. In fact, I don't believe I have ever had a U.S.-born Latinx professor. The number of classes I've attended with female mathematics professors hasn't been much better. I felt passionate that to create the change I wanted, I had to become a professor and be that change. Then, as I found out I had been accepted to SFSU and the CSU-LSAMP Bridge to the Doctorate program, I learned of another life changing event.

## Starting Graduate School and a Family

Prior to moving to San Francisco, I met the love of my life, Davi Pakter. I attribute my longevity in mathematics in part to his unwavering commitment and encouragement over the years. This support was certainly tested in my final semester of undergrad when we learned we were expecting our first child. It felt daunting to imagine simultaneously becoming a mother and beginning a master's program. Davi reassured me we were in it together and that I was capable of anything. I decided to accept the offer at SFSU and plan to welcome our first child, Ayla, just before Halloween during my first semester of the master's program.

To prepare for the fall I met another life-time mentor, Dr. Matthias Beck, who helped me redesign my schedule to better accommodate Lay's birth. Dr. Beck eventually became my master's thesis advisor and was influential in my choice of research areas. After a reading course on combinatorial topics (i.e., polytopes and Ehrhart theory) he suggested a research topic that eventually led to my master's thesis on Generalized Dedekind-Bernoulli Sums.

As my advisor, Dr. Beck respected that I was a mother while challenging me to become a better mathematician. We would often meet at cafés or parks where I could bring a stroller and let a curious babe explore the world while we explored mathematical arguments. There is a day in particular when, after sitting in an airport working out a proof and waiting for a return flight from a conference, I shared my argument for a particular theorem. I was both shocked and thrilled when a usually reserved Dr. Beck gave me a high-five. That day I felt like a mathematician.

A motto Dr. Beck shared with me one day, 'live your life and fit the career around it' has helped me reconcile my choice of being a mother first, then a mathematician. Still, I was unaware of the needs of a new mother and the minimal accommodations afforded to them in academia. I survived more than I thrived. Instead of postponing all my first semester course work to the spring, I continued working and even took a take-home midterm the day after giving birth. Over the following years, I relied on fellow grad students for childcare during classes, pumped milk in the bathroom stall, and had the baby and stroller in tow to study sessions. I struggled to stay confident as I balanced family with school and doubted my abilities and academic choices. It took a mental toll on me and I eventually sought professional therapy, which I have continued ever since. I also gathered advice from other female mathematicians, uncovering a belief that still persists today. To succeed as a mother and mathematician, expect to work harder, be demanded more of, and constantly have to sacrifice one for the other. In my

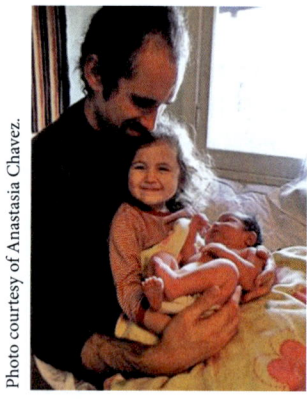

Photo courtesy of Anastasia Chavez.

Our growing family.

opinion, I see this as a response to a system built to promote a certain subset of the population exclusive of women, mothers, and people of color. Upon welcoming our second child, Ash, and recognizing the importance of showing all young girls the limitless futures available, I decided to pursue a doctorate in mathematics.

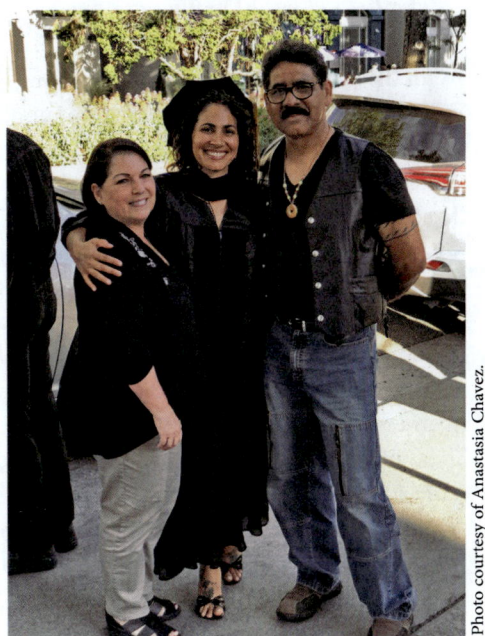

Photo courtesy of Anastasia Chavez.

"In mathematics you don't understand things. You get used to them." I came across this quote, attributed to John von Neumann, on a tea bag label my first year of graduate school at UC Berkeley. It is a quote that adorned my children's photo in my office for all of the six years I was there, and it is a quote I now share with my students on their first day of class. Dr. von Neumann's perspective exposes a truth about math: once the fear and unfamiliarity of it is removed, then all you have are toys and tools with which to play.

With my parents on my graduation day, 2017.

My experience shows, unfortunately, that keeping some folks afraid and unfamiliar with mathematics is how many wish to practice it. This is precisely the element of mathematics that I have never gotten used to.

I expected to fail my first attempt at the prelim, a six-hour written exam spanning two days that all mathematics doctoral students must pass before their fourth semester at UC Berkeley. Yet it was still surprising how difficult the exam was and how poorly I was prepared for it. Thus, before the semester even began, I was already sorted and stamped as deficient. I was one of a handful of Latinx graduate students and mistaken for the other brown woman in the department several times my first year. I was the only female graduate student with a family and one of very few to have attended community college. By the end of my first year I had all of this used as evidence for my inevitable failure in the department. A prominent visiting professor, who agreed to support my studies in preparation for a third attempt at the preliminary exam, told me I was not fit for research—I was "probably good for someone of [my] background," and that a PhD from UC Berkeley would not be worth the trouble. Just weeks later, after expressing my desperation of needing at least one ally in the department to believe in me, another professor said, "I do believe in you. I believe you could and should get your degree. I just don't think Berkeley is the right place for you." Messages like these, spoken by gatekeepers to the academy often who do not look like me, exemplify the inherent bias and discrimination built into academia originally meant to serve a select group of society. Although I walked away in tears from each of these gutting experiences, I was emboldened to earn my degree and use its leverage to unveil the inhumanity experienced by students, imposed by professors, and protected by the institutions that hire them.

# Mental Health and the Mathematics Community

With the support of my fellow students, my therapist, and the Disabled Students Services who fought for me to receive test accommodations, I eventually passed the preliminary exam. I had also sought advice from my future co-advisors, Dr. Lauren Williams and Dr. Federico Ardila. Their encouragement to plan for success, not failure, affirmed my resolve to succeed on my own terms. And still, much like during my first years of motherhood, I was fighting to survive as a mathematician. Soon after the prelim exam hurdle was jumped, I found myself depleted and deflated by the process. I felt my relationship with math had strained to the point of breaking, and I was ready to leave the program and perhaps mathematics all together. After talking to Dr. Ardila, who took my burnout seriously, he suggested I find a project that allowed me to reconnect with what I truly enjoyed about mathematics: curiosity, exploration, and discovery. From this grew my dissertation in the areas of algebraic combinatorics and matroid theory, which are at the heart of my research currently. Both Dr. Williams and Dr. Ardila were essential to my success at UC Berkeley. They valued my input, encouraged me to attend conferences and collaborate with others, understood and supported my choices when my family was involved, and made me feel seen as a human doing mathematics. When I asked them to help me honor my childrens' sacrifice over the years in support of my education, they joyfully awarded them home-made diplomas as all three of us walked across the stage during my graduation ceremony. The simple acknowledgment and validation of a person's humanity is paramount to a person's success. Sadly, it is glaringly absent all throughout our educational system.

I want to highlight the importance of personal well-being and emotional support in graduate school, and in all stages of life. Thus far, graduate school has been the biggest test of my self-confidence and commitment to achieving my goals. I reckoned with obvious discrimination and discouragement and I also became acutely aware of the personal effects these experiences have on underrepresented populations. My graduate experience opened my eyes to pervasive issues like stereotype threat, cultural obligation, being a spokesperson for a minority group, isolation, and impostor syndrome. I can recall the first time I learned about stereotype threat during a meeting with an older Black graduate student. I shared my sense of overwhelm as the only Latinx student in my classes; that my failure would be seen as a failure for all Latinx and all female students; that I couldn't help but worry that by failing I was only proving the nay-sayers right: people like me can't do math. These beliefs, along with the lack of encouragement and abundance of ridicule, affected my ability to be a productive, engaged, and valued

Photo by GradImages™

Youngest doctorates.

member of the math community for many years. Initially, I avoided the activities expected of a successful graduate student. I did not attend seminars. I declined most departmental events. I didn't discuss mathematics with faculty and rarely with other graduate students. With ongoing counseling and finding trusted friends and allies in mathematics, I was able to eventually mitigate the harm caused by these psychological and emotional phenomena. With depression affecting more than half of all graduate students, and suicide prevalent in our culture, I advocate for increased mental health services available to all and to continue removing the stigma attached to seeking support. In particular, among our faculty, this can be done by prioritizing student well-being and legitimizing student accommodations.

## Current Life and Words of Inspiration

I am now completing my last year as a postdoctoral fellow. I have held appointments at the Mathematical Sciences Research Institute and UC Davis under the mentorship of Dr. Jesús A. De Loera. And I'm proud to say a legacy of higher education has been made in my family. Both of my sisters are college graduates, and each are pursuing science-related careers. I've seen the math community acknowledge the realities of systemic racism and sexism and address the consequences with renewed drive. I am inspired by the graduate students of UC Davis who quickly respond to injustices by raising awareness to hold peo-

Photo courtesy of Anastasia Chavez.

Our family.

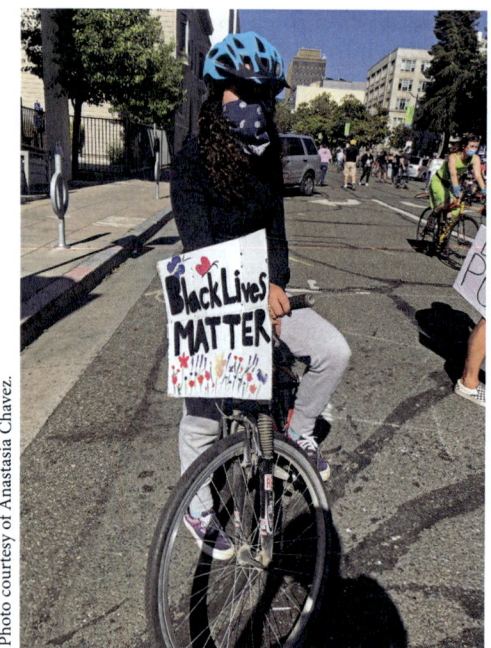

Photo courtesy of Anastasia Chavez.

Marching for BLM.

ple accountable. Lack of representation is being taken seriously by several institutions. Admission policies are changing and biased hiring practices are being rewritten to incentivize diversity along with research. Some institutions continue to approach these issues as just a quota problem, unaware that increasing the numbers alone cannot solve the problem. And still, without a greater presence of those affected by such commonplace discrimination, especially in positions of power, their voices will go unnoticed and silenced. We need policy reform at all stages of education, we must re-imagine the academic process, we must redesign the fundamental guiding principles of our educators, and we must fill the roles of mentors and advisors with compassionate, thoughtful, kind advocates who reflect the future of STEM.

As I consider the next step in my career, I am humbled by current movements that highlight the gravity of slowed or stalled progress. The countless murders of Black Americans and the incarceration of Black and brown youth weighs heavy and is too often paralyzing. Yet, when I see the courage, creativity, and ferocity of my own children as they march for women's rights, LGTBQ+ rights, and chant Black Lives Matter, I have hope that change is inevitable and our futures are brighter than ever.

# 6

# Dr. Minerva Cordero

## Early Life

I grew up in the countryside in Bayamón, Puerto Rico, in a neighborhood called Buena Vista. The place was beautiful, with rolling hills and plenty of space to play outdoors. I was the fourth of six children; I have three sisters and two brothers. My mother, Flora Braña Santana, was the ninth of ten children. Her parents did not allow her, nor her siblings, to attend school past the fifth grade because they were needed to help with the farm. She later became a seamstress. My father, Germán Cordero Rodríguez, was a tall, strong young man. After second grade his parents asked him to get a job to help the family and he was not allowed to return to school. He later became a truck driver

Dr. Minerva Cordero

and drove an 18-wheeler truck moving cargo throughout the island. Despite their lack of opportunity to get an education and the lack of financial resources, they emphasized the value of education. They saw it as the key to a better future. Growing up, my siblings and I liked going to school because it was the only outing we had most of the time.

School was the only place we had to learn about other countries and cultures. It was our way to leave our small town and "see" the world. My family never took a family vacation—indeed, when we would read stories at school about a family going on a vacation, we wondered what that must feel like. We were a family of eight, and we did not own a vehicle large enough to fit all of us. The only vehicle in the family was the one my father purchased and used during the day for public transportation. He would have a neighbor drive it six days a week, taking people from the neighborhood into town for a fee. On Sundays, his only day off from work, my dad would go out with his friends. When summertime came around, we were not very happy because this meant staying home all the time. The happiest day in the summer was the day before classes would start. This was a great holiday, especially for the girls. We would set up our school uniforms on top of one of our two beds, and all four sisters would sleep in the other bed together so the next morning we could all get dressed quickly for school. When I tell this story to my children, they tell me over and over, "that's not normal, mom."

**First school years**. Throughout elementary, middle, and high school, my teachers were very supportive and encouraging. I feel fortunate that I started my education by attending the neighborhood public school where we were all Hispanics (Puerto Ricans), and all came from a low socio-economic background. Our talents and determination distinguished us from each other and not the race or color of our skin, nor the size of our houses.

Photo courtesy of Minerva Cordero.

Minerva at ten years old.

From first grade to ninth grade, I attended the neighborhood school, called *Escuela Segunda Unidad de Buena Vista*; it later became *Escuela Segunda Unidad Cacique Majagua*. Almost all students (at least 90%) were provided free breakfast and lunch.

Moreover, we were given free shoes and a school bag at the beginning of the school year for several years. It was interesting that the shoes were the same for boys and girls—they were not very attractive! However, we needed them, so we were happy to get them. I had two excellent teachers in mathematics; from fourth to sixth grade, I had Ms. Solonida Hernández, and from seventh to ninth grade, I had Ms. Idalia Figueroa. They were smart, caring, and inspiring. My third-grade teacher, Ms. Sara Vargas, was the first to show me the beautiful and fun side of mathematics. Besides mathematics, I also loved my Spanish classes, and I developed a love for literature. My seventh to ninth-grade teacher was Ms. Elba Rodríguez; she was very passionate about the literary works that we would analyze in class. I looked forward to going to her class every day. Well, actually, I loved going to all my classes. I remember the joy of learning English with Ms. Marta Álvarez in seventh to ninth grade. I remember when she received a set of the SRA Reading Laboratory®, which was a program to learn how to read in English. It consisted of a set of cards with stories and questions to answer after reading the stories. It was designed by levels of different colors. She offered me to use it, and I was so excited that I went every day during lunchtime to take the tests. This was not required, but I enjoyed reading the stories and seeing how far I could go with the program. I was so thrilled every time I would advance to a new color. I now realize how much those cards helped me learn English.

From tenth to twelfth grade, I attended the high school closest to my home, which was called *Escuela Superior Miguel Meléndez Muñoz*. A quick look into the performance of the students at the school recently shows that it is on the top 50% of high schools in Puerto Rico, yet the percentage of students achieving proficiency in math is 35–39% (which is approximately equal to the Puerto Rico state average of 36%) for the 2016–17 school year. The percentage of students achieving proficiency in reading/language arts is 50–54% (which is higher than the Puerto Rico state average of 48%) for the 2016–17 school year.

Despite the overall school performance, I felt that it prepared me well for college in Puerto Rico. I especially liked my science teacher, Mr. José Manuel Erazo, who taught me

biology in tenth grade, and chemistry in eleventh grade. In eleventh grade, I took Algebra
II with Mr. Joselín Alonso, and that was when I truly fell in love with mathematics. Since
there were few subjects taught at the school, and I was close in age to my sisters Olga and
Lilliam, I took two classes with my sister Olga—Biology in tenth grade and Algebra II in
eleventh grade. With my sister Lilliam I took two classes too—Geometry in tenth grade
and Chemistry in eleventh grade.

With my baby brother, Germán, I took the Agriculture class when I was in twelfth
grade and he was in tenth grade. We were members of the Future Farmers of America
(FFA), and I loved participating in the events sponsored by the FFA at my school. One of
the events was public speaking. I participated in the local competition and won first place
at the statewide competition. My speech was titled "The Role of Women in Agriculture."
In my address, I included anecdotes of my mom and her two sisters working on the farm.
I recalled that after the Q&A session with the judges, one of them said to me: "you are
missing your true calling; you should be a lawyer because I assure you that you will not
lose a case." I guess I appeared pretty confident as I defended the role women had played
in agriculture, contributions that had not been recognized in the male-dominated culture
of agriculture. Looking back, I think how lucky I was that all my teachers looked just like
me during my first twelve years of school—they were my first role models.

**Road to mathematics**. For as long as I can remember, mathematics came effortlessly
to me. Having two older sisters who would often talk at the dinner table about what they
were learning in school motivated me to advance in school, especially in mathematics.
On the other hand, being the fourth of six children allowed me to help my two younger
siblings with their schoolwork. Helping them with mathematics came naturally to me.
One day, when I was in eighth grade, I attended a college algebra class with my cousin,
who was a sophomore in college. The instructor was introducing the quadratic formula,
which I had not studied before. The explanations seemed very logical to me, and I solved
all the problems he assigned the class. Before I realized it, my cousin and several of her
classmates were surrounding me, and I was explaining to them how to solve the problems.

A few years later, when I was in tenth grade, my uncle decided that he wanted to take
the General Educational Development (GED) test. However, since he had only gone to
school until the fifth grade, he needed to learn all the mathematics he would need to pass
the exam, so I offered to teach him. We met daily for several weeks and it was challenging
helping someone who had never encountered mathematical abstraction. He was curious
about many things, including how the variable '$x$' could have many different values. I
remember one day, after solving a problem whose solution was $x = 5$; he asked: "but before
you said $x$ was nine; how is that possible?" After many long hours working together, he
took the exam and passed it the first time. After this experience, I was convinced that
helping people learn and enjoy mathematics would be my lifelong endeavor.

At the end of my junior year in high school, I learned that there would be no math
or science class that I could take during my senior year due to teacher departures. So, I
decided to get a book that would help me prepare for the college entrance exam. Every day
after school and during the summer, I would sit outside on the balcony with a tiny blackboard and the book, and review all the different subjects covered in the exam. When the
exam results were announced at the school in November of my senior year, I was so happy

that I did well. Indeed, I received the highest score that anyone had received at my school up to that point. I remember my history teacher, Mr. González, excitedly telling me, "you are a monster."[1] And to this day my score remains the highest for the school after so many years.

## Undergraduate Education

I started college at the University of Puerto Rico's two-year campus in my hometown (called *Colegio Regional de Bayamón*, CRB), which was about an hour commute from home. I knew I liked mathematics the most of all subjects, so I decided to major in mathematics. However, I did not know the career choices for someone with a degree in mathematics, besides teaching mathematics in the K–12 setting. My oldest sister had learned about a new field of study people were going into those days: computer science. She explained to me all the benefits of pursuing a degree in computer science would offer. While I was tempted, and for two days changed my major to Computer Science, the day before classes started I decided my passion was mathematics, and I switched back to mathematics. The first math course I took in college was precalculus. Most of my science major classmates were registered in calculus because they had already had precalculus in high school, but I had not. So, I enrolled in the two-semester precalculus sequence. (I could have taken the one-semester five-hour course, but instead, I wanted to take my time with the basics. This is one piece of advice I give young people: "don't rush through the important things. Take your time to learn the basics, and later you can catch up," Which I did.)

**Tragedy hits home**. On August 9, 1978, just one week before my sophomore year in college would start, my family experienced a tremendous loss. On the evening of August 2, my father returned home from work and, after having dinner, went to do mechanical work on the truck he drove at the time, just like he had done many other times before. The truck belonged to his cousin, and my father worked for him as a driver for many years. He knew the truck very well; earlier, he noticed that there were some issues with the brakes. As usual, he would try to figure out and solve the problem himself. As he was working under the truck that was propped up with a regular car jack, the truck dropped on him, smashing his legs and carrying him for some distance down a small hill. My youngest brother was inside the truck assisting with shifting gears. My brother did not know how to drive at the time, and my father was not an experienced auto mechanic. After one week in the hospital, my father passed away, and our lives changed forever.

**University of Puerto Rico at Río Piedras**. After two years at the *Colegio Regional de Bayamón*, I transferred to the University of Puerto Rico's main campus in Río Piedras (UPR-RP). Getting to campus now meant approximately a two-hour commute in each direction because transportation from Buena Vista to Río Piedras was complicated and unreliable. Of course, we did not have the resources to buy a car. In my senior year, my sister Olga and I decided to rent a room in a house in Río Piedras. It made going to school so much easier. That year, we took a graduate topology course together. That was one of the best courses I took at the UPR-RP, and not just because my sister was in it!

---

[1] "*Eres una monstrua*" is a common expression in Puerto Rico when someone is excellent at a task or skill.

When I was a senior in college, my abstract algebra professor at the University of Puerto Rico, Dr. Carol Knighten, asked me about my plans for graduate school. At that time, I had not even considered graduate school. With my father passing when I was a sophomore in college, finishing college and starting a new job was necessary. Dr. Knighten told me, however, that I should consider graduate school and get a PhD because I would have better and higher-paying career options. There were two hurdles to cross to follow Dr. Knighten's suggestion: first, to pursue a PhD in mathematics, I would have had to leave Puerto Rico because there were no PhD programs in mathematics in Puerto Rico at the time. Second and equally important, I did not have the financial means to do so. However, Dr. Knighten recommended that I apply for a National Science Foundation Minority Graduate Fellowship, which I did. I received the fellowship and was able to take care of the financial hurdle.

Although my mother worried about me leaving, she was supportive of my plans to come to the United States to pursue a PhD in mathematics. Had Dr. Knighten not approached me about pursuing graduate school, I guess I would have never thought about it and would not have had the satisfying career I've had as a mathematician and an educator. Never underestimate the power of words of encouragement. I graduated from the University of Puerto Rico (UPR) with High Distinction.

**The Cordero sisters**. The oldest of my siblings is my sister Nilda. Nilda, who started college at the age of 16, obtained a bachelor's degree in biology from UPR after transferring from the CRB. She later completed a master's degree in Chemistry at the UPR and was a high school teacher at a public school in Bayamón for 30 years. Olga, who is two years younger than Nilda (and a year older than me), completed a bachelor's degree in mathematics at the UPR after transferring from CRB. After she finished her undergraduate studies, she worked for an engineering/architecture firm in San Juan, Puerto Rico.

The year I decided to come to the U.S. to pursue graduate school, Olga decided to follow and went to the University of Iowa with a Graduate Fellowship provided by the university. She transferred from Iowa after her master's degree and received a PhD in Statistics from Utah State University. My youngest sister, Lilliam, who is one year younger than me, studied computer science at the CRB. By the time Lilliam started college, the two-year institution had become a four-year college.

As I mentioned earlier, my parents did not even complete elementary school, yet all four girls studied STEM careers. Besides our love for STEM, we also loved literature, art, and music. We would talk for hours about all the things we were learning in those subjects at school. During my senior year in college, my teacher in literature thought I was a Spanish literature major and was quite disappointed when she learned I was not pursuing a degree in literature. Unlike the girls in the family, my brothers pursued interests outside of STEM. My oldest brother went to vocational school (and studied woodworking), and my youngest brother studied business in college.

# Graduate Education

I started my graduate studies at the University of California at Berkeley. It was a big cultural surprise moving from the little town in Puerto Rico where I lived with my family to Berkeley, California, where I would live with roommates. The biggest shock was the university environment. It seems that all the graduate students at Berkeley came from very top universities from all over the world and knew so much more math than I did at the time. I felt an extreme sense of estrangement and alienation, and the demands of my studies, plus the challenges of living in a place so different from where I came from, made it hard to connect with people.

While in Puerto Rico, I would often visit my professors during office hours. At Berkeley, I felt very intimated by them, not just because of the language barrier but also because they seemed intimidating. I was fortunate that my first year I took Complex Analysis from a new visiting professor who had just joined the department, Dr. Gustavo Ponce, who was originally from Venezuela. I thoroughly enjoyed his class and decided to take other courses in complex analysis. The next semester I had a course with Dr. Donald Sarason, who later became my master's thesis advisor.

After completing my master's, I moved to the University of Iowa where my sister Olga was studying mathematics. When I moved to Iowa, I felt more comfortable talking to my professors.In my first semester, I took an abstract algebra class with Professor Frank Kosier.He was very entertaining in class and seemed very approachable.When discussing a theorem or concept, I noticed that sometimes he would write the letters "BS" next to it. Being the careful note-taker I was (and not knowing what it meant), I would also "designate" some results and concepts as such. After a couple of weeks of lectures, I decided to go to his office for some clarifications. I started by saying how I was familiar with most of the theorems and concepts we were studying but just had a question about why for some results he added the letters "BS" next to them, and for some others, he did not. He looked at me puzzled, paused for a moment, and then blasted into laughter. He then said: "Minerva, if you don't know what *BS* stands for, I'm afraid I can't tell you what it is." When I shared this experience with a couple of classmates, they laughed and finally told me what it meant. I felt so naïve.

The second semester, Professor Norman Johnson was my abstract algebra instructor. One day I was in his office during office hours, and the writings on the board caught my attention. There were many geometrical figures all interconnected. After he explained his research, I was fascinated by the subject and read several papers he recommended. I wasn't sure how to approach him about conducting research for my doctoral dissertation under his guidance. Luckily, after achieving a near-perfect score in the doctoral written exam in algebra, he approached me and asked me to work with him for my dissertation. Since his research was at the intersection of algebra, combinatorics, and geometry—three subjects I loved—it was very easy for me to agree to work with him.

**Teaching mathematics in college**. At the University of Iowa, I was given the opportunity to teach as a graduate teaching assistant. I was terrified at first of standing in front of a room full of undergraduate students and having to teach in English. I remember one day in class, one of my students asked a question. I did not quite get what he was asking, so I asked him to repeat his question. After a couple of times of me asking and then not understanding his question, he said, "Oh, never mind." I asked him to repeat what he said, and very frustrated, he said again, "never mind." I had no idea at the time what "never mind" meant. Instead of interpreting it as a single phrase meaning 'don't worry about it,' I had understood it as two separate words. Fortunately, my office-mate helped me understand the intention the student might have had.

Despite these issues initially, I immensely enjoyed being a TA, and in my classes, there was always laughter and good camaraderie among my students. I tried to share with my students the difficulties I was experiencing as a student myself in a 'foreign' country with a different language. I found the students to be quite receptive and encouraging. A comical situation I experienced when I was a graduate teaching assistant at the University of Iowa in charge of teaching a section of Calculus I, was when I was discussing how to use the first and second derivative of a function to draw its graph. I did an example and could tell that the students did not grasp the ideas well. So, I decided to do another example. After I completed that example, I asked the class if they would like to see another example, and they said yes. As I started to work on my third example, I could hear the students talking in the back, and a few were giggling. I turned around and asked what was going on. One shy young lady said, "I think you're doing the very same example for the third time." It was hilarious.

**Tragedy strikes again**. My mom was one of the kindest, most positive people I have ever met. When she was little, she was one of the few children in her class to have shoes, and since all the kids walked to school with no shoes, she would hide her shoes before getting to school and go shoe-less with them. When she was 54 years old, she felt something on her chest and went to the neighborhood clinic doctor. The doctor told her she was fine and dismissed her. Two months later, she could still feel it, so she went back. And, at 54 years of age, she had her first mammography test. It turned out she had stage IV metastatic breast cancer.

Minerva and her mother Flora.

After the diagnosis, she underwent radiation and chemotherapy, but it was too late; a year and a half later, she passed away. Sadly, I attribute her misdiagnosis and untimely death to the dismal quality of healthcare available to the working poor.

We buried her on December 31. I was still in graduate school and it was so hard to find the motivation to keep going with everything. She had been my strongest supporter, and receiving my PhD was going to be my greatest gift to her. But now, she would not see me graduate. I went into a deep sadness and was not interested in doing anything, not even mathematics.

That year I was living in Bowling Green, Ohio, where my then-husband had a visiting professorship. In February, he was giving a talk at the University of Michigan and I decided to join him on the trip to Ann Arbor. There was a terrible winter storm, and the roads were covered with snow. On the way home, we had a terrible car accident. I remember laying there in the snow, thinking I was going to join my mom.

Due to the injuries to my ribs and pelvic bone, I was unable to walk for three months following the accident. I would just lay in bed and think about dying and being with her. I felt so much pain during that time, but all that pain, all that sacrifice, was also the motivation to finish my PhD I did not want all to be in vain, even if my mother could not see me graduate in person. I knew wherever she was, she would be watching over me, and she would be proud.

**Doctoral dissertation defense**. The first talk I ever gave was my doctoral dissertation defense. As an undergraduate or graduate student, I did not have the opportunity to attend a conference or give a talk. So, for my defense, I practiced so many times what I was going to say and write (at the time, the presentation was given using a blackboard with no visuals) that I memorized every single word I would say. However, at one point in the middle of my defense, I noticed that the audience seemed uneasy, so I asked if there were any questions. My supervisor quickly replied: perhaps you would not mind repeating the last result and its proof but in English this time. I did not realize it, but I switched to Spanish in the middle of my dissertation without even being aware of it.

# Research—Finite Geometries

My research area is at the intersection of algebra, geometry, and combinatorics. It is called finite geometries. Specifically, I study finite projective planes coordinatized by non-associative division algebras; these are called semifields. These finite geometries belong to the larger class of combinatorial (balanced incomplete block) designs.

In my work, I classified and discovered several new classes of semifields. Such an algebraic structure is somewhat less than a field in the sense that multiplication may well fail to obey the associative law. Still, the corresponding projective plane has a fairly rich collineation group. Notice that other "weakening" of the algebraic structure called a field have been widely investigated over the years, both from an algebraic and a geometric point of view. For example, nearfields fail to be fields if just one of the two distributive laws is missing. While the work of Zassenhaus in 1935 classified finite nearfields, there is no complete classification for finite semifields yet. In my research, I always combine algebraic methods with geometry. So, in my work, groups, rings, and polynomials always come together with collineations, subplanes, particular configurations of points and lines, etc. The challenge is always how to translate any piece of algebraic information into geometric data.

One of the research results I am most proud of is a conjecture I made relatively early on in my career about the existence of a particular class of semifields. A few years later, a group of young mathematicians in Italy proved the conjecture. Currently, I study other finite algebraic combinatorial structures, including loops, quasigroups, and nearfields. These structures are used in algorithms to encrypt and decrypt data in a safer way than using associative structures.

# Professional Career

The day after my doctoral dissertation defense, I flew to California for my first job interview. I was so nervous, worrying that I might revert to Spanish in the middle of my talk, but I did not. It was an enjoyable experience. I later interviewed at another school in California and one in New York. I received two job offers for tenure-track positions in mathematics departments. However, I did not accept either position and instead moved to Lubbock, Texas, where my husband had a faculty position at Texas Tech University. There were no positions for me, but like many women with a PhD, I moved to the place where my husband had a job.

Since there were no faculty positions at Texas Tech when I moved to Lubbock, I decided to contact the school district to see if I could teach in one of their high schools. As it turned out, they needed a math teacher for Estacado High School, and I was offered a position, which I accepted without ever visiting the campus. At the time, Estacado High School was the lowest performing school in the entire school district. The campus had over 95% minority enrollment and over 90% economically disadvantaged students. Unfortunately, the school suffered from a high turnover of teachers, which, as research shows, has a very negative impact on student success. During the in-service training, I met the principal and other teachers at the school. The principal was very careful in emphasizing how much we needed to show love and concern for the students.

I was very thrilled that these were things that the school valued. Unfortunately, I was not prepared for the environment in the classroom. I loved mathematics and was so excited to teach this group of students. I understood they needed someone to teach them mathematics and show them that they cared for them; I treated my students as my friends. Very soon, unfortunately, I found myself in a very chaotic classroom due to my ignorance regarding high school effective classroom management skills.

In addition, one day, at the end of class a young girl asked me if I would take her home with me because her situation at home was challenging. I went to the principal to ask what the right thing would be to do at the time. He instructed me not to take the student home with me. The next day as I was finishing class, the young girl started a fight with another girl. She threw her classmate on the floor and was hitting her. Students started screaming, and a teacher from the next-door classroom called the police and they interfered. After everyone left the classroom, I sat there and burst into tears. I realized that what I wanted to do the most, which was to teach these children mathematics and show them that I care about them and their future, I was simply not good at because I did not know effective classroom management techniques. I walked to the principal's office and told him that I was not helping these children that desperately needed help and that I had decided that that day would be my last day at the school.

For the next several months, I immersed myself in learning the ins and outs of publishing in mathematics, and I submitted a couple of manuscripts. Around that time, Texas Tech had an opening for a faculty position for which I applied. After my talk, one of the faculty members told me that my talk was the best he had ever heard in his 30-plus years at Texas Tech; it made me feel content. I received an offer and started my first tenure-track position the following fall.

## Personal Life

I have a wonderful family that values faith and family above all. My husband, James Alvarez,[2] is also a mathematician and a colleague at the University of Texas at Arlington.

---

[2] Dr. James Alvarez is featured in Chapter 1.

 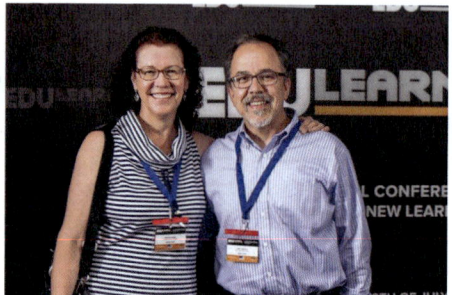

(L) Minerva's family at her son's Alex graduation from Yale (2014). (R) Minerva with Dr. Álvarez at an education conference in Spain. Photos courtesy of Minerva Cordero.

We have two sons. Alex is a composer and lives in New York City. Our youngest, Nicholas, is a college student majoring in mathematics. Both of our sons are accomplished violinists, and when they were in school, we always had music playing at our home. We are very close and enjoy family vacations which are primarily around visiting our extended families in Puerto Rico and San Antonio.

I am very close to my three sisters, and we talk on the phone often. Three of my siblings live in Puerto Rico (Nilda, Lilliam, and Wilfredo, my oldest brother). My sister Olga lives in France, and regularly teaches at the American University in Washington, DC. My youngest brother, Germán, with whom I was very close, passed away in 2010. Now that I have been in academia for some time, have achieved full professorship, and have been working as an associate dean in the College of Science at the University of Texas at Arlington for six years, I feel that my efforts to increase the number of women and Latinxs who pursue careers in science and mathematics are more fruitful.

## Reflections on Diversity

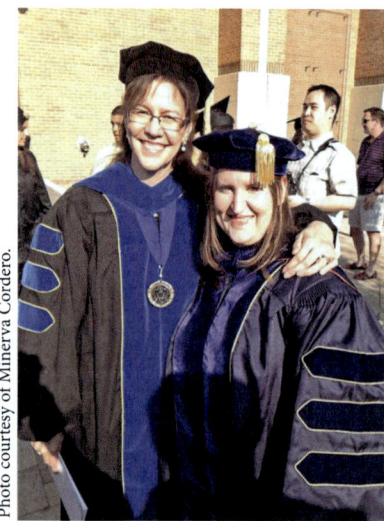

Photo courtesy of Minerva Cordero.

Minerva with her first doctoral student, Dr. Angie Brown.

Growing up in Puerto Rico and attending school at the University of Puerto Rico has been something that definitely helped me learn mathematics. I never felt that I was less qualified or less capable of learning mathematics than my peers because I was a Hispanic woman. Indeed, both in high school and in college, I received the highest award for a graduating senior in mathematics. My sister Olga, who is just one year older than me, also received these awards. It took some time after I came to the United States to realize that people could question my ability to do mathematics because I was a woman of Hispanic origin.

I remember the first year as an assistant professor, a colleague told me I got my position because I played the Hispanic woman card. I had no idea what he meant by that. His explanation did not make sense to me. The conversation ended with him telling me that I was lucky because I was the "good kind" of Hispanic, because "I didn't look it." It was not until several years later that I understood the hurtful nature of his words. I could not help but feel sad for my fellow Hispanics and women colleagues like me who have dealt with this mentality all their lives. I wish I could say that I have seen the mathematics community improve, but that has not been the case.

Minerva at the graduation of Louis Stokes Alliances for Minority Participation Bridge to Doctorate (LSAMP-BD) Fellows.

Last year, I was selected as an IF/THEN American Association for the Advancement of Science (AAAS) ambassador. This program aims to increase the number of girls interested in careers in Science, Technology, Engineering, and Mathematics (STEM). For several years now, I have been involved in outreach activities to bring more underrepresented minorities and women into mathematics. I have given many talks to groups of fifth- and sixth-grade girls and their parents. Often, the parents only speak Spanish, and I feel very fortunate that I can talk to them and explain to them all the opportunities that their daughters have if they get interested at a young age in science, engineering, and mathematics.

## A View Towards the Future

More Latinxs/Hispanics should get involved in mathematics research and mathematics education. Having a diverse group of individuals working on this will help attract a more

Minerva at SACNAS Leadership Institute Reciting an Afro-Caribbean Poem.

diverse group of people into mathematics. With the changing demographics of this country and Latinos making up a larger percentage of the population, it will be imperative that we are participating in mathematics. To encourage young people in math, I show them the power and the value of mathematics.

Without mathematics, we would not have the advances in technology that exist today. For example, their cell phones would not work as well as they do if someone had not spent the time doing the mathematics required. Without mathematics, we would not have the advances in medicine that exist today. We could not even see movies and pictures so clearly as we do today without all the advances in mathematics.

Hispanics are the fastest-growing and youngest group in the U.S.—expected to comprise 30% of the U.S. by 2040. If we do not make efforts to attract Latinx students into mathematics and other STEM areas, the U.S. will not be able to maintain its current position in the technology world.

My advice to students is to take as much math in high school as possible, but not move too quickly. It is essential to learn the basics very well before moving on. With a strong background, it is easy to catch up and finish a college degree on time. The important thing is to persevere and not give up, remembering that mathematics is for everyone. You don't have to be a straight-A student in high school or college to have a successful career as a mathematician. That's the beauty of it.

# 7

# Dr. Ricardo Cortez

## The Beginning

My story is about family, education, and service.
My parents, Beatriz and Jaime, grew up in the
same neighborhood in El Salvador. They were both
from relatively small and low-income families.
They knew each other from an early age and began
dating as teenagers. When my mother finished
high school, she pursued a job opportunity as
an executive secretary at the United Nations and
moved to New York City at the age of 19. My
father, who is infinitely resourceful, found a way
to follow and moved to New York shortly thereafter without knowing how to speak English and
without knowing how he might support himself
there. That was the beginning of a decade in New
York full of momentous occasions including their
wedding, the birth of their first son, my brother

Dr. Ricardo Cortez

Illustration created by Ana Valle.

Jaime and two years later the birth of their second son, me. The family was complete six
years later when my sister, Beatriz, was born back in El Salvador. My brother has been
a musician all of his life and is now a nationally recognized liturgical musician based in
Arizona. My sister is an award-winning artist and a scholar of Central American literature
at Cal State Northridge. My parents have been married for 58 years now. You will hear
more about them later in the chapter.

In New York, we lived in a multicultural neighborhood in Astoria, where the next-door
neighbors were Greek and others were Italian. I only lived there for four and a half years
and did not attend any school. My brother attended public school for a year, which he enjoyed, but not everything was reassuring to my parents. There were rumors of candy laced
with drugs being given to children and other problems that, combined with the reality
that my parents could not afford any private school, led them to relocate the family back
to El Salvador. There, as a two-income family and substantial sacrifice, my parents could
afford to send the kids to a Jesuit school, *Externado de San José*, where I did all 13 years of
primary and secondary school.

# Growing Up

The Jesuit School was rigorous academically and focused on social justice. There was a "social hours" requirement every year that could be fulfilled by volunteering at any one of a list of places kept by the school. I remember participating in food and clothing drives after storm-induced landslides devastated entire neighborhoods.

I was generally a good student, but I was not interested in every subject and I didn't make my best effort when I was not interested. So, my grades were good in some subjects and acceptable in others. I was in seventh grade when I first discovered that I was good at mathematics. It wasn't from positive reinforcement, but from a day in which the teacher did not agree with my conduct in class. On that day, two classmates and I were tossing a ball around the room every time the teacher had his back turned. When we were caught, he singled me out for being disruptive and said sarcastically that I shouldn't worry because he didn't give low grades based on conduct. A few minutes later he retaliated by writing a problem on the board and pointing at me to come up and do it in front of the class. I don't remember the problem, but I did it right and, more importantly, I knew I had done it right before the teacher validated it.

In high school, I had very good math teachers who encouraged my interest in mathematics and made me feel important. One of my memories from that time is about a day I went to ask my algebra teacher questions about a procedure we learned in class. I distinctly remember his answer starting with "I am going to tell you something because I know you will understand." This made me feel like I was being entrusted with information that the teacher considered beyond the scope of the class, but he knew I was capable of understanding. It was a simple statement but it had a huge impact on me. I think of that statement and choose words carefully when I talk to students now because I don't know which statements will resonate with them.

My high school cohort was organized as a community. We arranged for one or two of us to lead tutoring sessions in different subjects for the benefit of the other students. I was selected to lead the mathematics sessions and attended sections in chemistry as a student. Outside academics, I was interested in soccer and classical guitar. I was good at soccer and played with the school team for most of my years there. Classical guitar was a different story because it was clear right away that my brother was much better at it and I wasn't going to be in the same category as him. So I took it slowly and played only as a hobby.

Discrimination exists everywhere, even in places with little racial diversity, where the victims are those with darker skin tones and the poor. We were taught to care for the less fortunate and advocate for those who don't have a voice or are not heard. How to do this in practice was not clear. Food drives were important and necessary at critical times, but their effect was momentary and didn't produce long-term benefits. The political environment in El Salvador reached a boiling point after decades of oppression of the poor (which was the majority of the country) by a series of military governments. A war broke out when I was a teenager and life became difficult. I knew people who were killed or physically disabled from bullets. Childhood friends left the country on a moment's notice. I knew the smell of teargas and the anxiety of having to get home before curfew. All of us did.

A family decision was made that when my older brother and I finished high school we would move to the United States to go to college. We were lucky because we were U.S.

citizens and could travel freely. Still, the reality was that my family could not afford to send two kids to college. Nevertheless, I was one school year behind my brother, so he went first.

Nobody in my immediate family had a university degree except one of my uncles who was an architect. Certainly, no one in my family knew a scientist or artists who made a living as artists, so when my brother mentioned that he liked music and I mentioned I liked mathematics, the advice was that we study something that my parents viewed as a career, like engineering, rather than a science or an art, which they only knew as a subject matter without a realistic option of a career. My parents' position was that we could study anything we wanted as long as we had engineering to fall back on.

## College

My older brother left El Salvador and moved to the United States as a freshman majoring in engineering at Arizona State University (ASU). A few years earlier Jaime spent nine weeks in Arizona as an exchange student and he maintained close contact with the host family, the Langstons, who became an extension to our family and remain so to this day. Having them nearby was a great relief to my parents and the number one reason for choosing ASU. A year later, when it was my turn to move out, I also went to ASU and began my studies in mechanical engineering. Moving to Arizona was a cultural shock in every way. I had just turned 18 and went from living with my parents in San Salvador to the dorms in Tempe. I did not know how the educational system worked, how the social system worked, and, while I thought I spoke English, I didn't understand half of what people said to me. When I did understand, it took time for me to find the words to answer. My personality was different without the vocabulary to tell jokes or say something clever. I also felt some pressure to contribute financially to my education before depleting my parents' savings.

I do not remember anyone who spoke Spanish in the dorms where I lived with my brother. The only exception was a father and son team from Mexico, who worked at the dorm cafeteria cleaning and washing dishes. After speaking with them a few times I found out that I could work at the cafeteria for minimum wage, which was not a lot, but if I worked the breakfast and dinner shifts I could get my food for free. So that's what I did to earn some money during my first semester in college. On weekends, I had a second job as a referee for young kid's soccer games. I had a bicycle that got me to the different locations and I had a lot of fun refereeing. After a few months, my English was getting better and my studies were going well. I knew there was University-sponsored tutoring in mathematics and I thought perhaps I could do that so I applied for the job. I started tutoring calculus while I was still a freshman and I kept adding more tutoring courses to my repertoire every semester. I worked as a tutor with that University program until I graduated.

I earned enough at my jobs to pay for my living expenses excluding tuition. I had inherited some of my father's resourcefulness, and I found out that there were many scholarships and grants opportunities that I qualified for. Some of them provided a couple of hundred dollars, others were a little larger. Every year I would apply for a handful of scholarships in hopes I would be awarded some. This system worked well for the rest of my time at ASU without having to take on any loans.

My engineering advisor, Dr. Davidson, was helpful. He always took the time to talk with me and to work with my ideas. The school of engineering offered its own mathematics courses "for engineers." When I told him that I preferred to take the equivalent courses from the mathematics department, he approved every one. A conversation with him that changed my academic path without my realizing it took place when I had run out of mathematics courses in the engineering program. I told him that I liked mathematics and that I wanted to continue taking one math class every semester even if they were not required. He said that if I was serious about it, that I should add mathematics as a second major. I didn't know that one could do that, but I thought it was a great idea. I became a double major that day.

## Family and Education

I did not finish my bachelor's degrees simultaneously. I finished my mathematics degree first and had a few courses left to take in mechanical engineering. My brother Jaime switched majors from engineering to music, which was his passion in the first place, and graduated a year after I did. A few years later, my younger sister, who was born and raised in El Salvador was unable to make steady progress in her studies due to the political instability in the country. In a moment's notice, she moved to Arizona, where my brother lived, and continued her education at ASU. She earned a PhD in Literature and Cultural Studies. A few years later, as a professor at Cal State Northridge, she found the time to attend CalArts and complete a master's in Fine Arts. Now she has two concurrent careers: art, which is her passion, and the faculty position where she inspires students in the Department of Central American and Transborder Studies.

When my brother and his wife Kari started a family, my parents didn't want to miss the chance to be a part of the lives of their grandchildren and decided to also make a move from El Salvador to Arizona. They were in their early fifties and did not know what type of jobs might be available to them, but family was most important. As it turns out they were able to find jobs and had time to enjoy the grandchildren. My mother worked as an interpreter in the courts of the city of Mesa, where she was able to take classes to earn a particular certification.

More importantly, taking college courses revived her desire to earn a university degree, which had been on hold for years. She started taking courses at the local college and sometimes at ASU. My father, who didn't want to feel left out, decided to take some courses as well. They were motivated to finish and declared majors in justice studies (Mom) and studio art (Dad). Their three kids had degrees from ASU and now it was their turn. My parents graduated from ASU together in 2004.

Photo courtesy of Ricardo Cortez.

Mom and Dad's graduation.

# Graduate School

Many people have similar stories about how they ended up in graduate school. For me, the path was not straight, mostly because I did not know what people did in graduate school or what the future held for somebody with a graduate degree. But some of my ASU math friends were better informed and applied to graduate programs. Among them was Stella, who is from Los Alamos where her father was a scientist. She went to the University of California (UC) Berkeley for graduate school in mathematics. During her first semester there, when I stayed behind finishing engineering, I communicated with her and she encouraged me to apply to Berkeley. She said a lot of great things about it; I trusted her. So I decided to apply to the mathematics graduate program at Berkeley; in case that didn't work, I started interviewing for engineering jobs. That was the master plan. Besides the graduate application to Berkeley, I sent one to Arizona State University because my advisor suggested I do that.

I didn't know much about graduate programs or if a master's degree was a prerequisite for a PhD degree. I knew Berkeley was a top-tier institution. I was worried about my academic preparation. I decided to apply to the master's program with the intention of switching to the PhD program after a year. I don't know if I discussed this issue with anyone or if I did it on my own. To this day, I believe that decision made it possible for me to get a PhD. I wasn't ready for a rigorous program. It was absolutely necessary for me to spend my first year of graduate school taking upper-division undergraduate courses.

There is a story about the way I found out I was admitted. During spring break, I still had not heard back from Berkeley. So, I decided to take a 13-hour road trip and visit the campus. I walked around the hallways, listened to conversations, and tried to get an idea of the environment. I went to the math office, and they sent me to the graduate coordinator. He asked the secretary for my application, but she couldn't find it so, he looked on his desk and found it among a bunch of other applications. He picked it up, opened it, and said: "you are admitted."

When I walked out of the graduate coordinator's office, I crossed paths with a Latino student, but we did not exchange words. By an unbelievable coincidence, he was a PhD student from El Salvador, whose father had been a teacher at the Jesuit school that I attended. Herbert Medina found out that I was an incoming student and he sent me copies of past preliminary exams for me to study in advance. Since then, Herbert has been a source of support over the years to me and hundreds of other students. He is now Acting President at the University of Portland.

In Berkeley, I became more aware of the problems faced by underrepresented minorities, especially as they relate to access to education. It was around this time that I learned about SACNAS, the Society for Advancement of Chicanos and Native Americans in Science, and I attended

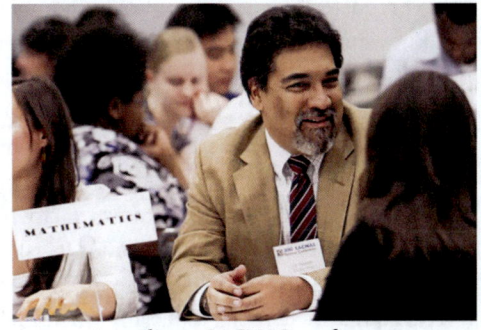

Attending a SACNAS conference.
Photo courtesy of SACNAS (Society for Advancement of Chicanos/Hispanics and Native Americans in Science) sacnas.org.

their conference for the first time. It was a relatively small meeting held in January, and I really liked the idea of providing opportunities for undergraduate students to learn about graduate school and careers in scientific research.

I became involved in SACNAS by attending the annual meetings and eventually becoming a student board member. I did not miss many SACNAS conferences for the next 25 years and worked along with many others to increase and enhance the mathematics component of the conference. Over the years, we organized mathematics sessions at the conference and created a task force that wrote proposals to fund the participation of mathematics students. Later, we were able to connect the mathematical sciences research institutes to SACNAS so that more members of underrepresented minorities could learn about the institutes and perhaps participate in their programs in the future.

As a graduate student I was already interested in education and had the opportunity to work for programs that tried to make a difference in the education of students who were marginalized and didn't have advocates. I got connected to these programs through friends who were in graduate programs in mathematics education. At the time, I was inexperienced and didn't know how to determine the likelihood of success of these intervention programs or if they were designed based on best practices or research. Looking back I can see that some of those programs had little chance of success when, for example, they hired early graduate students like me to teach a course at the local high school (with no training) in order to correct a systemic problem like tracking, where students of color are often placed in remedial mathematics courses, which eventually limits their opportunities in college. At the time I didn't have the maturity to understand that I was spending a large portion of my time on activities that were noble but had little chance of effecting lasting change. Today, I would strongly advise students to dedicate the academic year to their PhD studies and get involved in equity work only during part of the summer. Otherwise, the risk of falling behind in their studies is too great.

My time in the PhD program was full of nerve-racking moments when I needed supporters to make it through. My entire first year was an example of one. This was because, at the time, the UC Berkeley mathematics department admitted many more students than they expected to complete the program. The justification was that by doing so, they gave opportunities to students who wouldn't usually have them. However, specific support systems for all of us to succeed were difficult to find. As a result, most of the students in my cohort dropped out of the program in the first year. The transition from undergraduate to graduate student required redesigning my study habits, learning to read textbooks, and knowing when to keep trying to solve a problem and when to ask for guidance. I took the first-year exams twice without success. There was a process to petition for a third try, which I did and passed the exams. I had made the transition.

Another nerve-racking moment was asking a professor to be my dissertation advisor. I had taken two courses with Alexandre Chorin, and I decided that I would like to work under his supervision. I didn't know exactly how to ask or if he would accept, so it took me a few days to finally go to his office to ask him if he would be my advisor. His reply was "I was hoping that would be the case." I immediately knew I was in good hands. He was a great source of support, especially when I felt discouraged or questioned my ability to finish. When I graduated, I received an award for an outstanding dissertation in applied mathematics.

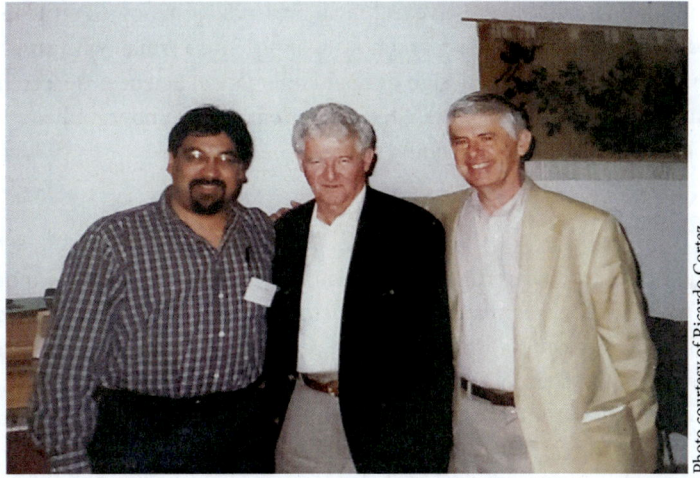

Dr. Cortez (left) with Peter Lax (center) and Alexandre Chorin (right).

## The Summer Math Institute

One program that had a huge impact on me was the Summer Math Institute at Berkeley. It was created by Uri Treisman and mathematics professor Leon Henkin specifically to address the underrepresentation of people of color in mathematics. Black, Latinx, and Native American undergraduates spent six to eight weeks at UC Berkeley taking a fast-paced course in an area of advanced mathematics for a couple of weeks before transitioning to group research projects. I was a graduate student at the time and worked as a Teaching Assistant in this program for two summers. The mathematics, as well as the engagement in research, were new to the students. The program had an administrator making sure that students and instructors had everything they needed to do their work and also organizing social events.

The students were divided into two groups of 12 where each group worked on a different mathematical topic. The program also included outside speakers and professional

The MSRI-UP 2012 cohort.

development seminars on how to give technical presentations, how to write technical papers, what graduate school is like, and how to apply. This format, with some modifications, became the model for the design of the Mathematical Sciences Research Institute Undergraduate Program (MSRI-UP), a research program for undergraduates that I co-created in 2007, and that is still going strong.

## Postdoctoral and Faculty Years

Near the end of my graduate program, I applied for a National Science Foundation grant to go to NYU for three years as a postdoctoral researcher. I also applied for jobs in other places, including the National Security Agency which hires the largest number of mathematicians in the country. During a phone interview, they said that due to security risks, their employees had to sever ties with people from a list of countries that included El Salvador. I pointed out that it was impossible for me and others with family in Latin America to comply, and therefore, their policy was discriminatory toward Latinxs. I was not invited to the next interview. A decade later, I found out that the policy was still in place when a graduate student at Tulane University was forced to give up an NSA fellowship because his girlfriend was from a Latin American country on the list.

I got the National Science Foundation fellowship, which brought me back to New York, closing a circular path in my life. Besides being a time of professional growth, it was in New York where I met Kathy, who was in charge of grants administration at the Courant Institute at NYU. Kathy and I moved to New Orleans to work at Tulane University and have been together ever since. She has been a selfless supporter throughout the years and brightens every day of my life.

## Final Thoughts

Working toward systemic changes involving deeply rooted  practices requires actions from many angles and a substantial investment of energy and time. Even when institutions are willing to change, they call on the few people of color to do much of the groundwork. This disproportionate request to do equity work has led some to decline such invita-

Mom, Dad, siblings, Kathy, Kari, and nephews Daniel, Nicolas and Benjamin.

tions. While change happens at a frustratingly slow pace, there are two things that I try to remember. One is that this is a lifelong endeavor most likely to be characterized by incremental changes before major breakthroughs can happen. The second one is that to be effective in this work one has to be successful by the current measures of the system. For this reason it is important to recognize key moments when one must focus on professional advancement and self-promotion in order to reach positions of influence where one's efforts can be more effective in the long term.

I came to understand this during my postdoctoral years, which I dedicated to establishing a research program that would lead to a good academic job and extend through the transition to faculty member. As a professor, I made the decision to dedicate time to work toward increasing the participation of people of color in mathematics at the cost of a reduction in research publications and other professional output. This was my personal decision and it is not a recommendation. It required constant assessment to make sure that I advanced professionally. As an Assistant Professor I declined invitations to lead undergraduate research programs until after tenure. Instead, I involved small groups of undergraduate students of color in projects connected to my own research that would produce publishable results. I was lucky to have the mentorship of members of the SACNAS community, and especially of my colleague, long-time collaborator and friend, Lisa Fauci. Their support was critical to overcome setbacks and any obstacles placed in front of me.

# 8

# Dr. Jesús A. De Loera Herrera

This is my story. I present it through my answers to three questions: *Where did I come from?*, *Where am I today?*, and *Where am I going now?*.

## Where Did I Come From?

I was born and raised in Mexico City, the huge city of twenty million stories and passions. I come from a middle-class Mexican family. My nuclear family was a bit unusual because my parents separated when I was young, so I was mostly raised by my mother Antonia Herrera Tejada and my grandmother Margarita Tejada de Herrera. My sister Judith is just three years younger than me. The four of us, and a bunch of dogs, formed my household.

Illustration created by Ana Valle.

Dr. Jesús A. De Loera Herrera

I grew up on the edge of *colonia*[1] Narvarte and went to elementary school at the adjacent, rough neighborhood of Buenos Aires. Because of this I faced some difficult experiences: my lunch was stolen, I was beaten on a couple of occasions by gangs, and the police harassed young people like me. These experiences made me aware of poverty, violence, and social injustice. Maybe because of the harsh reality I was living, books were my true friends: I could read and read all day long, or at least until my mother ordered me to go outside.

My father Jesús De Loera López was a remarkable man. He was a poor farmer with no more than a junior high education, who went from being a *bracero*[2] during the 1940s to becoming a congressman and a gubernatorial candidate in Mexico. Sadly, my relationship with my father was broken. I was proud of his accomplishments, how he pulled himself up from nothing, and I wanted to emulate him, but we spent such little time together that he gave me nothing to hold onto.

My mother is an even more remarkable person to me! With little more than a secretarial degree and divorced from my father, she raised two children. She supported us, believed in all of our dreams, no matter how stupid or insane they were, and she sacrificed everything for us. She and my grandmother gave me lots of love, guidance, and a calm environment to grow.

---

[1] A *colonia* is a residential quarter within a city.

[2] A *bracero* is a migrant farm worker.

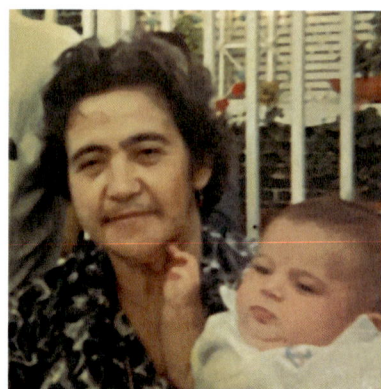

Photos courtesy of Jesús A. De Loera.

My family and me.

I went to public schools in Mexico City. From very early on, I was always a diligent student, because I loved school, I was excited to learn, especially history. The principal of my elementary school had been a historian and he took a liking to me. He gave me books and encouragement. My mom really promoted learning as a way to improve oneself. At great expense and effort, my mom would take my sister and me to the pyramids and excellent museums around the country, which I loved.

When you work in math, people often treat you as gifted, but I do not think of myself in that way. I am actually quite slow to understand things, but I firmly believe that anyone that really loves something enough to try to be good at it, will become good at it. The very first time I remember loving mathematics was in middle school. My teacher asked us to carry a daily mathematics diary where we would complete our homework. When grading these, she gave prizes to clear, well-organized answers with explicit reasoning. I remember spending so much time making sure my answers were neat. I loved the introduction to basic axiomatic geometry proofs, using similar triangles and parallel lines. It was so much fun to give a solid argument! To know the truth!

After a successful national exam I was admitted to a high school associated to the National University of Mexico (UNAM). *Preparatoria* 6 is in the bohemian neighborhood of Coyoacán, very close to the house of Frida Kahlo and Diego Rivera and Trotsky's last home. Me and my friends took advantage of the intellectual and artistic atmosphere of the area. I fell in love with astronomy and I helped to build telescopes at the *Sociedad Astronómica de México*, I wrote poetry and read the work of philosophers from Plato to Nietszche, and I learned to speak proficient French. I was blessed with a very rich intellectual atmosphere among my classmates too, who were also "coming of age" as thinkers and creators. Many of my classmates are now academics.

In high school my fascination with mathematics grew and I first became aware that I was fairly competent when I won third place in a high school mathematics competition. I loved scholarly and theoretical pursuits, but labs were not as exciting to me, so when it came time to decide for a college major I decided to study mathematics. Admission in UNAM was done by scores and grades: since I had the highest scores, I could have chosen medicine or law as my major, but I chose math because I loved it. Many people told me I was crazy not to choose a more conventional career.

My high school friends.

In retrospect, it was indeed an impulsive decision, purely based on a feeling. I had no idea what a mathematician did, no idea of what professional math really was about when I chose that career. Today, I can say with certainty that mathematics as a career is an excellent choice, with plenty of job choices and financial rewards too, but back then I did not stop to think about money and what I wanted in a job. All I knew was nothing gave me so much pleasure as mathematics did. Fortunately my teachers gave me reassurances and my mother never once doubted my choice.

In 1984, I went to college at the *Facultad de Ciencias* of UNAM. This is a huge university, with more than 100,000 students in one campus. It offered me unlimited opportunities, but you had to push hard to be noticed. It was very hard. As a freshman, for the first time in my life, I felt really academically under-prepared. All of my classmates were so much smarter than me. Being the first to go to college from my father's side of the family, I certainly felt like an impostor. But I persisted. How did I become resilient? It is hard to say, I feel it is mostly an inner fight. The Aztecs said, "*un gran guerrero no es quien logra dominar a sus enemigos, sino aquel que puede dominarse a si mismo.*"[3]

During my university studies, I found that math was even more beautiful than I ever imagined! I truly loved the subject, and I could work until late hours of the night thinking about and enjoying the challenges from class, even when it was hard work. I quickly learned that mathematics is not a spectator sport! You learn math by doing math. Just like in music or sports, practice makes perfect.

I was lucky to have brilliant classmates that made creating mathematics even more fun and exciting, like playing a soccer match! We would argue about theorems and proofs or about life and politics, until late into the night. In yet another lucky life event, I connected to four mathematicians who trained me and believed in me: In analytic geometry class I met Javier Bracho, an elegant geometer topologist who promoted imagination and color-

---

[3] This quote translates to: "a great warrior is not the one who manages to dominate his enemies, but the one who can dominate himself."

Photo courtesy of Jesús A. De Loera.

My college mentors.

ful results. Later Francisco Larrión, an algebraist who took me under his mentorship and taught me to write mathematics and be rigorous. Victor Neumann, one of the founding fathers and elders of discrete mathematics in Mexico took me under his wing. I was his teaching assistant for graph theory and combinatorics and I discovered I enjoy teaching. Victor introduced me to Gilberto Calvillo, the first real applied mathematician I ever met, who at the time worked at the Bank of Mexico. He taught me about operations research and mathematical economics and I discovered that math is crucial to solving real concrete problems. Math became a power tool for analysis and decision-making, not just a beautiful creation. The artistic side of mathematics mingles really well with its applicable power.

The reality is, I am the mathematician I am today because these men were true mentors: they gave me a lot of their time, constructive criticism, and encouragement. I recognize today the value of being a mentor and a supporter of young talent. My 1989 senior thesis was in combinatorial topology and group theory. I gave a modern proof of the classification of planar Cayley graphs of finite groups, first presented by Matschke in 1899. That was the first time I heard about *polyhedra*, my favorite mathematical objects. In two dimensions, these are high school polygons. Cubes, crystals, and pyramids are examples in three dimensions.

In perhaps the luckiest event in my entire life, I met a truly brilliant and lovely physics student Ingrid Brust-Mascher, who was not only the top student in her class, but was to become my best friend and later my wife. Taking a risk, Ingrid and I left Mexico together in Fall 1989. In 1990 we got married, and attended graduate school at Cornell University in New York together. Ingrid was an applied physics PhD student, while I worked on my PhD in applied mathematics with a minor in operations research. Ithaca was a drastic change from Mexico City, but a welcome change for us as newlyweds. The life in the outdoors and true winters was a lovely new experience after the hectic life of the big city.

I did not know it at the time, but the late 1980s brought a remarkable group of mathematicians to Cornell. The Center for Applied Mathematics (CAM) was comprised of a diverse collection of researchers and graduate students whose work covered all areas of research. The director, John Guckenheimer, promoted new ideas and creativity. Lou Billera,

Photos courtesy of Jesús A. De Loera.

With Ingrid in 1990 (left) and in 1995 (right).

who had made great contributions to mathematical optimization, game theory, and algebraic combinatorics gave me wise guidance. Mike Stillman, co-inventor of the computer algebra system *Macaulay* and one of the pioneers in making algebraic geometry computational, also influenced my way of thinking about mathematics. At the time of my arrival to Cornell several prominent Russian mathematicians, including Andrei Zelevinsky, Misha Kapranov, Sasha Barvinok, had visiting positions in Ithaca. The atmosphere was stimulating and engaging. In my first year I found the best PhD advisor I could dream of, Bernd Sturmfels, then a young rising star in the field of computational and applied algebraic geometry. Since then he has created a whole movement around computation in algebraic geometry (computing with systems of polynomials equations and inequalities). Bernd was my most important teacher. He believed in me and became a good friend. Once more a key mentor helped me improve and grow.

In those days I saw polyhedra appear everywhere, in applied mathematics, e.g., optimization and probability, and even in the context of pure math (algebraic geometry and topology). Polyhedra became the emphasis of my PhD dissertation and in fact my entire career. In my PhD work, I solved an open question of Gelfand, Kapranov, and Zelevinsky by finding an example of a non-regular triangulation of the Cartesian product of two simplices. I proudly used a computer-based proof. I also wrote a couple of papers on algebraic algorithms for manipulating systems of polynomials. To this day, this topic continues to fascinate me. I received my PhD April 25th, 1995. That same year in June, our first son Antonio was born. By the end of August, we drove across the country to take jobs at the University of Minnesota, two fresh PhDs with a young baby.

My job was at the Geometry Center. There, with the emphasis on computers and geometry, my research style matured. While at Minnesota, Victor Reiner helped me to explore our common interest on geometric combinatorics. Our second son Andrés was born in Minneapolis, at a hospital by the Mississippi river.

After Minnesota, my second job was at the Computer Science Department in ETH-Zürich Switzerland. I was hosted by the research group of Emo Welzl and Juergen Richter-Gebert, both great friends who had a deep influence on me. The research atmosphere was creative and joyful. Those were happy times for me and my small new family. It was

Photos courtesy of Jesús A. De Loera.

With Bernd Sturmfels in 1996 and with family in 1999.

exciting to work on various problems in convex and discrete geometry and computational geometry while, on the weekends, I could escape with my children to the Swiss Alps.

While in Zürich I worked on the problem of finding optimal triangulations and subdivisions of polyhedra. I developed practical algorithms to find triangulations with the fewest number of simplices. For example, I discovered that the minimum triangulation of the regular dodecahedron has 23 tetrahedra. In 1999, I moved to the University of California, Davis to take on a tenure-track faculty position.

## Where Am I Today?

As I write this recollection of my life I have already completed 20 years in the faculty! It has been a long personal and intellectual journey.

My work in combinatorics and discrete geometry, started as a PhD student, continues. My first book, written with my great friends Joerg Rambau and Francisco Santos, *Triangulations: Structures for Algorithms and Applications*, was published in 2010. It is a thorough reference on triangulations of polyhedra. By now, my scholarly work touches on several other topics.

I have made noteworthy contributions to the problems of computing volumes and integrals over polyhedral regions, and counting lattice points. These three computational problems have many applications, from pure math (algebraic geometry and representation theory) to applied combinatorics, probability and statistics (one is the analysis of contingency tables, see Figure 8.1). The software project *LattE* was started under my direction and initiative, with the purpose of carrying out those computations. *LattE* is used by many mathematicians and it helped to introduce dozens of students to research.

In the past ten years, my desire to work on more applications and computations has led me to apply algebra and geometry in the area of *combinatorial optimization*. This is the part of applied mathematics related to making optimal choices. A famous example of such problems is the *traveling salesman problem*: Given $n$ cities that must be visited only once, and the costs $c_{ij}$ of flying from city $i$ to city $j$, the goal is to find the best order to visit all of the given cities in order to minimize the total cost of the trip. I am proud to have been one of the leaders of a new approach to the theory of combinatorial optimization. In this new point of view, we use tools from algebra, topology, and geometry, that were previously

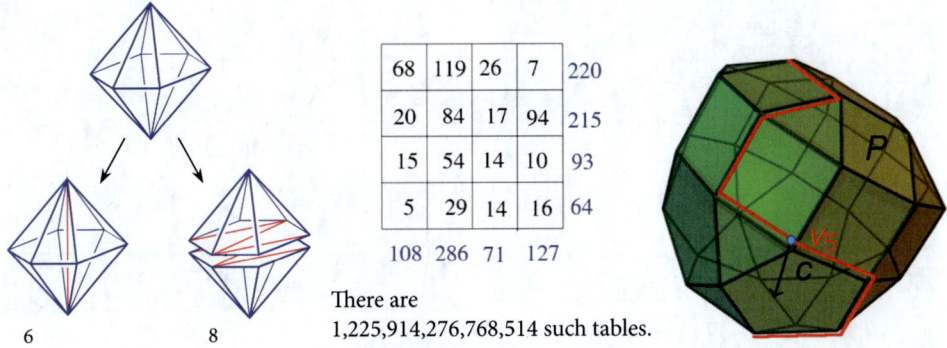

|     |     |     |     |     |
|-----|-----|-----|-----|-----|
| 68  | 119 | 26  | 7   | 220 |
| 20  | 84  | 17  | 94  | 215 |
| 15  | 54  | 14  | 10  | 93  |
| 5   | 29  | 14  | 16  | 64  |
| 108 | 286 | 71  | 127 |     |

There are
1,225,914,276,768,514 such tables.

**Figure 8.1.** (L) Two triangulations of a polytope. (C) Counting statistical tables. (R) A path of the simplex method.

considered too pure and unrelated to applications, to prove unexpected computational or structural results. For instance, a long-standing geometric question I care about asks to bound the diameter in the graph of a polyhedron, that is, the maximum length of a shortest path between a pair of vertices in its graph (see Figure 8.1). This is relevant to understanding the performance of the simplex algorithm, one of the most influential computer algorithms in history. My second book, coauthored with Raymond Hemmecke and Matthias Köppe, is titled *Algebraic and Geometric Ideas in the Theory of Discrete Optimization*. This book is the first compilation of results from this geometric perspective.

I have also written many papers on convexity, such as my papers about variations of Carathéodory, Helly and Tverberg type theorems. I like theorems with colors! Tverberg's theorem is one of my favorites: Suppose $a_1, \ldots, a_n$ are points in $\mathbb{R}^d$. If the number of points, is sufficiently large, namely $n > (d+1)(m-1)$, then they can always be colored into $m$ color classes $A_1, \ldots, A_m$ in such a way that the $m$ convex hulls conv$A_1, \ldots,$ conv$A_m$ have a point in common.

All of these years of work I have been blessed with appreciation and recognition. For my contributions to discrete geometry, optimization, algebraic algorithms, and mathematical software, I was elected a fellow of the American Mathematical Society in 2014 and as a fellow of the Society of Industrial and Applied Mathematics in 2019. Through mathematics I had unique experiences, including traveling all over the world to conferences, discussing my work with some of the brightest minds, and being invited by the Obama White House to speak at the State Department.

Still, the most lasting reward is all of the students I have mentored. I have now worked with over 60 undergraduates on projects. It is rewarding to see them grow. Some of them are now professors themselves. My 14 former PhD students have gone on to make me very proud. I even had the pleasure of meeting some of my "academic grandchildren." Of course, seeing my own biological children grow and prosper is also a great source of my life's joy!

## Where Am I Going Now?

I still have hope that I can create more mathematics in my remaining years and I am keeping up with new advances. Today, computers are essential to the discovery of results,

Photos courtesy of Jesús A. De Loera.

The people I have met are the true rewards of my career.

even in pure mathematics, but I see new opportunities as we enter a new era of computational mathematics. First, new methods, that rely on formal logic, make mathematical assertions automatically verifiable with higher certainty. It is not only the computer carrying on calculations, but also the logical components of the proof itself, including all the background theories. A second major shift is how data science and artificial intelligence are driving new mathematical questions. Finally computers and technology (e.g., online courses, Zoom) continue to change mathematics education and collaboration. I will focus my energy to think about such research.

If you have read this far you now know my story and it should be fairly obvious that my journey was only possible because I found people willing to support me. I did not arrive here all by myself. I would like to dedicate time to make mathematics accessible to everyone, especially young Latinx people who can later take my place in the circle of mathematical creation. The most important contribution I will ever make is the people I helped along their journey.

The rising Latinx population indicates it is in everyone's best economic interest to make sure Latinxs can have access to mathematics and science. As I write this *testimonio*, the world is engulfed in a pandemic, that has uncovered the deficiencies and the inequality that exists in our society. Thus, fighting for equity and inclusion in STEM is one of my most important duties.

## Parting Words

Dear young reader, if you have fallen in love with math, I urge you never to give up in your passion. Let that love guide your education to become the mathematician you dream to be. Do not give up in the hard moments of disappointment. Rest, but do not desist, persist! Always remember that an expert is someone who has failed enough times to understand how to avoid pitfalls. There are teachers and other students who are willing to help. There is nothing wrong with asking for help and guidance. Seek the mentorship and company of people that will support you and care for your dreams. Then, one day, you will become the expert, the teacher, the mentor. Take that place proudly and be generous and humble. Remember where you came from. Your journey was only possible because of others that came before you.

I wish you a healthy, balanced, and wise life. One where you have the time to think deeply about the world and others.

# 9

# Dr. Jessica M. Deshler

## Family and Identity

My mother, Clara, was the third of four siblings and the first in her family to go to college. She enrolled at the University of Albuquerque (no longer in existence) right after high school, earned her associate's degree in nursing and began working immediately. My father was from Texas, worked for the National Weather Service, and had been working for a short while in New Mexico. Soon after they married, they relocated to Northern Texas in the middle of 'tornado alley' where there was much work for the National Weather Service to do. The first few years there included my birth, a

Dr. Jessica Deshler

Illustration created by Ana Valle.

terrible tornado that leveled most of our city and a motorcycle accident that left my father with a traumatic brain injury. After his accident and subsequent rehabilitation, he was unable to go back to work, and moved in with his parents while my mother and I returned to New Mexico. I spent my childhood traveling back and forth between the two states to visit both sides of my family.

Most people don't realize I'm Hispanic when they meet me—I am white passing, I don't speak with an accent, my maiden name is Scottish and my married name is German. My father was white, and though I identify with my mother's side of the family, I pass as white and am therefore not the target of direct racism and acknowledge the privilege that this has afforded me during my life. Despite the privilege that comes with fair skin, it can also bring struggles of not quite fitting in or feeling like you can't quite claim your heritage because of how you look. Being mixed-race and white passing is a frustrating state to live in, but I like to believe it has helped equip me with tools to understand and interrogate issues of identity and culture.

## Education and Falling in Love with Mathematics

Growing up in Albuquerque I spent a lot of time with my grandparents and the rest of my mother's large family, who taught me how to speak Spanish. My mother worked hard as a nurse and instilled in me a strong work ethic that I maintain to this day. She was involved

91

Photo courtesy of Jessica Deshler.

Me with my Grandparents, Tranquilino and Adelicia Martinez, who were also my Godparents on the day of my baptism in 1977, Albuquerque, NM.

in my schooling: from helping with homework to being an active member of the PTA[1] throughout my education. School was always important to her, and she ensured I took it seriously and excelled. In middle school, I had an opportunity to take a class once a week from a visiting instructor. This course introduced a small group of students to some mathematical ideas we had not seen before (I specifically remember learning about functions and how we could label variables with whatever name or symbol we wanted. It was mesmerizing!). This was the first time I recall ever learning something about mathematics that was just *fun*.

I continued to join science clubs and enjoy mathematics and science courses throughout middle and high school. During my freshman year I was placed into a mathematics course with an amazing teacher, Mr. Martin Paco, who I was lucky to have for more classes later. He showed us, mainly through his goofiness, how much he cared about us learning the content. He walked around during exams answering questions, while also writing hints on flash cards and taping them to his back so we could see them as he walked around. He cared more about us learning than he did about how well we performed. I had one of his classes before lunch one year and recall many days that I skipped lunch because some concept or skill we were learning eluded me and I spent my lunch hour struggling through the mathematics in his classroom with classmates and with his help. The frustration was real, but so was the satisfaction of acquiring the skill or understanding the concept. When I took an Advanced Placement (AP) Calculus course from him later in school, he used to hold study groups at his home with his family. His wife prepared food, his kids were running around, and he helped us study for the AP exam. This has always been one of my most memorable high school experiences and arguably, he is the teacher that set me on a course of mathematical exploration throughout my life. The support he gave me worked and I became the first female student from my high school to score a perfect 5 on the Calculus AB AP exam.

I was a sophomore in high school when I realized how much I loved the struggle of mathematics, and the seemingly clear 'answer' at the end of a problem. I understand now the great complexity of mathematics, but in my early years I was attracted to what felt like the binary nature of getting a 'right' or 'wrong' answer to a problem. Despite this love of mathematics, I was encouraged to consider pursuing an engineering program in college. I even spent a semester during high school doing a student internship with a graduate student in civil engineering. It seemed that engineering would be a lucrative career and that appealed to others. Luckily for me, I'm stubborn and the work of that particular student was joyless enough to convince me that mathematics was the right

---

[1] PTA is the acronym for Parent-Teacher Association.

path for me. I also knew that going to a college out of state was not an option for financial reasons. I ended up enrolling in the school my family wanted me to attend—New Mexico Institute of Mining and Technology (New Mexico Tech). At New Mexico Tech, about an hour south of Albuquerque in Socorro, New Mexico, I took as many mathematics courses as I could. While I felt comfortable in most of my mathematics courses, some were very difficult, and some of the professors were intimidating (even if they weren't trying to be). To this day, there are professors from my undergraduate institution that make me stop in my tracks when I see them at conferences. I learned a lot from my professors, much of it during office hours and tutoring sessions, not just in the classroom. As a senior I had to choose a sequence of courses to take and I chose

Me with my mother Clara Garcia, ca. 1987.

*Photo courtesy of Jessica Deshler.*

courses on differential equations. I fell in love with differential equations during that year and continued to study them in graduate school.

I finished the requirements for my bachelor's degree in mathematics a semester early because I took some general education courses in the summers. Since I was finishing degree requirements early, my university informed me that my scholarship would end in that final semester so I would graduate in three and one-half years. While this was an achievement to be proud of, it left me unsure of my next step. I was graduating college in December and hadn't made plans. During the previous year, though, my department needed undergraduates to lead calculus labs and tutor in the learning center. I was hired as an undergraduate Teaching Assistant (TA) and taught a calculus lab for freshman. I absolutely fell in love with teaching that year and began to think about a future in academia. I had never had any interest in teaching in the public school system, but teaching calculus was amazing (though I'm sure I actually taught very little in that lab). I decided I would pursue graduate school and hoped my experience as a TA would help me get into a program and funded as a Graduate Teaching Assistant. I had decided to get both a master's degree and a PhD from a single institution with a notion that I could get to know the faculty while working on the master's before agreeing to work with an advisor on PhD research. The department where I earned my bachelor's degree did not have a PhD program at that time, so I knew I would have to go elsewhere. I also knew that staying close to home was still the frugal option and so enrolled at the University of New Mexico in my home city for graduate school.

I started graduate school in January and was greatly prepared for courses in differential equations, but greatly unprepared for courses in other fields. I was also unprepared for the drastic difference in expected workload between undergraduate and graduate courses. I don't recall if my undergraduate professors ever tried to talk to me about graduate school and the expected workload and performance, but certainly my family couldn't prepare me for it. They did support me, though. My first semester was rough, I joined a cohort of students who were halfway through their first year, I didn't have an orientation because they didn't do that for the Spring semester, and I had no exposure to numerical analysis at the undergraduate level, but was enrolled in a graduate level numerical analysis course. I ended up dropping below full-time status that semester as I dropped the numerical analysis course I knew I would not pass. However, I used the following summer to study and prepare myself for the next fall and became diligent in my study hours and in figuring out how to be a graduate student and do well in courses.

What I learned throughout all of my years in school was that nobody would know what was best for me as well I would. That nobody could know my experiences and my background just by looking at me.

## Work-life Balance and Moving Slowly

I started my graduate program directly out of my undergraduate program, and two years later when I was completing my master's degree, I decided to get married and start a family. I knew I wanted to have (lots of) children. I was not willing to wait until I was out of school because I knew that would take years. I had heard stories from faculty members who waited until they were in a tenure-track job, then waited until they earned tenure, then promotion,… and then it was too late to start a family, or it was difficult to do so. I had always wanted a large family. I made a decision early on that my family would be my focus and school would have to work around that. This was the beginning of putting my family and personal needs above my research and academic needs, or at least at the same level.

I was initially drawn to applied mathematics, and my first exposure to research was in this field, working with mathematics and engineering faculty members on research we conducted in a fluids lab on campus. We studied a particular type of fluid flow through experimental set-ups and numerical simulations. We did this for a year or so, then this project provided a tremendous opportunity in the form of a summer internship running the numerical simulations at a national laboratory using their software and computing facilities. While this was a great chance to get to know what industry research is like, it was also a chance for me to realize that working in a cubicle on a computer all day was not the environment where I would thrive. After the internship, and some more time spent collecting data in the lab, my advisor left my department for a position at another school. He had two doctoral students at the time—a single man with no strong ties to the city and me. At this time, I was already married with a couple of kids. The other student went with our advisor to his new institution and continued to work with him while I stayed in my hometown and floundered a bit. I stopped working in the program full time, I stayed home with my kids for a couple of semesters while I figured out if a PhD was really what I wanted. Eventually, I decided to meet with the graduate program director to determine my options. I probably should have done this sooner, but didn't know that this person was a resource for me.

At the time my advisor left, I was pregnant. While I knew that starting a family would slow down my degree progress, I also knew it was the right decision for me. The opinion that graduate school was the right time for starting a family was not shared by everyone around me. The graduate program director made it very clear to me during that meeting that he did not believe I should be pursuing a PhD while having children. (I wrote an article about this experience that many people seemed to be able to relate to.[2]) This interaction fueled me forward to completing my degree through choosing a different advisor and new research project. This time, I decided I wanted to know more about teaching and learning. A faculty member in my department, Dr. Kristin Umland, had just changed her research area from mathematics to mathematics education and she agreed to supervise me to write my dissertation.

I firmly believe I completed my degree just to spite faculty members who shared the opinion that I shouldn't be there, that the dissertation should be all consuming and the only thing on which I should spend time and energy. I've carried the drive I needed to complete the degree into my role as a faculty member. Not only do I still focus on family and integrate both my personal and professional lives as much as possible (my kids have been to LOTS of conferences in lots of cool places, and even lived overseas for a year so I could work there), but I also use my position to act as a role model for students. Graduate students still struggle sometimes with deciding whether they can have families while in school. I try to support them as best I can, and I am now more able to do so since I oversee the graduate program and all graduate students in my department. I am particularly focused on supporting women in mathematics and whatever choices they might make about their personal lives, in whatever ways I can. I ended up choosing a dissertation advisor during the end of my time in my PhD program who could relate to my personal circumstances. She had children slightly older than mine, and our research sessions sometimes included her children keeping my children entertained so we could work. I will be forever grateful for the work she did with me when we were both relatively new to mathematics education, and the work she did after I graduated as an advocate for me and for mathematics education. In my case, in particular, a faculty member in my (former) graduate department decided (after my graduation) that he did not believe mathematics faculty members should supervise dissertations in mathematics education and attempted to have my degree withdrawn by the university. Luckily for me, my advisor fought on my behalf and only relayed the story to me after it had been resolved.

Besides having children, I had other family commitments that occasionally took me away from research. My husband had to have a major operation while I was in school so I had to take a semester of medical leave to care for him. I also struggled academically sometimes. I already mentioned having to drop a course I wasn't ready for, but I also didn't pass all of my preliminary exams the first time I took them. These are only a few of the setbacks I encountered while in my program. I use them as examples of how real life can cause you to need to adjust your expectations. For me, these slowed down my success, but they did not stop it. Through all of these experiences I realized I needed to make decisions that were best for me, not for others, and that nobody would know what that was except for me. After receiving my degree, I took a faculty position nearly 2,000 miles from

[2] J. Deshler (2017). Mixing Babies and Graduate School, *MAA Focus*, Vol. 37, No.1. digitaleditions.walsworthprintgroup.com/publication/?m=7656&i=392392&p=0&ver=html5

Photo courtesy of Jessica Deshler.

Me with my grandfather, Tranquilino Martinez, at my PhD graduation in 2008.

home. That part of my journey is a common one that students considering academia need to be ready for—you go where the jobs are. In my case, the job was great, but it was very far from family. I anticipate relocating closer to home when the time feels right, but for now I will stay where I am for my children's sake.

## Research and Service

I have been fortunate to be able to pursue research in fields that inspire me. In particular, my experiences in graduate school have led me to believe that positive experiences in teaching and mentoring while in graduate school have a lasting impact on the future careers of mathematics graduate students. Together, my advisor and I navigated our way through the muddy waters of doing research in a field in which nobody else in our department worked and it turned out to be the best decision—it led me to my career in research in undergraduate mathematics education (RUME). I study the teaching and learning of mathematics at the college level.

I primarily study graduate students and their experiences. I love to work with them as they progress through their graduate programs and gain teaching experience and examine how their teaching practices and philosophies change over time. Professional development of this group, the next generation of mathematics faculty members, has become increasingly important as our students and our teaching environments change over time and I've been fortunate to be part of a group of people across the country studying this population and finding ways to support them and their students. I study how graduate students progress in their teaching philosophies and teaching practices as they participate in various teaching and mentoring experiences.

Photo courtesy of Jessica Deshler.

My kids in 2015 at the Liberty Bridge in Budapest, Hungary.

I am increasingly interested in the professional world of mathematics and the structural barriers it contains that work against the success of underrepresented mathematicians, including women mathematicians. I've studied programs that support women faculty members in mathematics, curriculum that provides opportunities for women in classrooms to have agency in their learning environments and programs we've built to support underrepresented students in calculus. I helped implement an Emerging Scholars Program (modeled off the work of Uri Treisman[3]) in our calculus sequence and have spent the last few years examining which aspects of the program help build community among our underrepresented calculus students and support their persistence in STEM. My current institution, West Virginia University (WVU), is a predominantly white institution (PWI) located in Appalachia and is a drastically different cultural environment than those of the Hispanic Serving Institutions (HSIs) that I attended as a student. Similar to how being white passing has required me to acquire a different set of skills, so has working in a PWI that has a very low percentage of underrepresented students. Developing programs to support marginalized students in this environment requires thinking differently about recruitment, logistics, and implementation. I've been fortunate to be able to bring funding to the university from professional societies and federal agencies to support some of this work.

Most of my research has been collaborative, which is fairly common in education work. I enjoy working with other researchers in mathematics education, but also those who study academic development, social sciences, education in other science disciplines and those who study the K–12 system. Collaborative work is an amazing way to learn about the world around you and to learn from others. I have been lucky to find ways to integrate

---

[3] Uri Treisman is a University Distinguished Teaching Professor, professor of mathematics, and professor of public affairs at The University of Texas at Austin. He is known for starting the Emerging Scholars Program, which works to ensure that all students, regardless of their life circumstances, have access to an excellent education. This program has been replicated in universities throughout the United States.

my research areas into my service, teaching and administrative work. I currently serve as the Graduate Program Director and Graduate Teaching Assistant (GTA) coordinator in my department, overseeing the advising and progress of approximately 50 full-time graduate students, the development of the graduate program and the professional growth of approximately 30 GTAs. I see myself as a role model, advocate and resource for women in mathematics. Besides serving on department committees, I have served as a faculty associate for the WVU Center for Women's and Gender Studies, as a Provost's Fellow in the Office of Graduate Education and on my university's Council for Women's Concerns. However, one of my greatest professional achievements has to be when I was selected as a U.S. Fulbright Scholar to spend a year providing professional development to mathematics doctoral students in Hungary. I have recently been promoted to Full Professor within my department.I am only the third woman to achieve this rank in my department's history, and the first Hispanic faculty member to do so. I see my job as one of making the process better for those who come after me, and for opening doors and creating spaces of safety and equity for those who are not always welcomed into mathematics.

## Personal Advice

As noted, I have four children who are always the 'WHY' for anything I'm doing. They are why I work hard, why I play hard, the reason I take care of business. It has been a joy watching them grow up and become their own people. They call West Virginia home, and have had a vastly different childhood than I did, but I hope they grow up believing it was a good one. When I was on sabbatical as a Fulbright Scholar in Hungary, my kids not only spent a year living abroad but had the chance to travel, as travel within Europe is inexpensive. We have visited places we had only read about in books previously. The pictures of my kids show them in amazing places around the world. These opportunities only happened because I didn't give up when I hit obstacles, and I have realized that if I never try (to get research published, to get projects funded, to apply to programs like Fulbright), then I will never succeed at them. I no longer wait for 'sure bets', I pursue what I want and know that eventually, some of it will happen. I seek out people who are doing the work I want to do, and having the experiences I want to have, and get advice. Do the same. You don't have to do this alone. Find people to be your support network. Don't give up, seek out mentors, ask for help, and find your own path, in your own time.

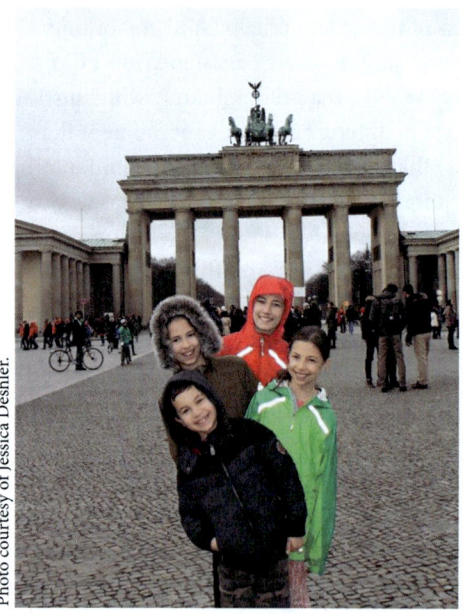

Photo courtesy of Jessica Deshler.

My kids in Berlin, Germany in front of the Brandenburg Gate, March 2016 during the Berlin Seminar (annual European Fulbright event).

# 10

# Dr. Carrie Diaz Eaton

## Early Life

I grew up straddling worlds. I was born and raised in Warwick, RI, a couple miles from Providence. My mom was also born in Warwick, RI, but my dad was born about 3700 miles away in Lima, Peru.

My dad, Hugo Evaristo Díaz Mendizábal, was the second oldest of nine children. He would tell a story about one day when they had no money to buy food and his mom went to the butcher. She placed a stack of butcher paper on the scale and demanded that she was owed that much meat over the years of buying from the butcher's store—and it worked. This story is indicative of the strong Mendizábal women, and also of the reasons that my father immigrated to the U.S. in 1960. As the second oldest, he and his older brother were the

Dr. Carrie Diaz Eaton

Illustration created by Ana Valle.

first two children to immigrate to the United States after his dad, a surveyor, immigrated and then promised them jobs with him. At fifteen years old, after completing his cartography certification at the *Instituto Geográphica Nacional* and speaking little English, he arrived in California—to a broken promise and no job.

My mom came from a working-class, New England Protestant family. My grandfather worked his way up to a foreman in a machine shop, and my grandmother was a homemaker. They built the house they lived in—living in the basement until the first floor was complete and so forth. They both survived the Depression, which forever left an impression on their lives. My grandfather enrolled in the Civilian Conservation Corps, a federal work program initiated by Roosevelt. They had an urban homesteader garden before it was trendy and when it was necessary. My grandmother could make dinner out of any scraps and passed her knowledge of pie making and preserve making to me.

My dad's employment prospects brought him eventually to the east coast. It was in Rhode Island in 1966 when he met my mom, on an invitation from his Filipino coworker to dinner at his wife's house. My mom, Becky Ober, was the wife's younger sister. In 1967, my mom graduated high school, married my dad, and was pregnant with their first child.

Photo courtesy of Carrie Diaz Eaton.

Becky Ober and Hugo Diaz dating.

I think I was aware of how unlikely my parents' (and my aunt's) marriage was. I was raised in my mother's church, but my dad went to a Spanish-speaking Catholic church in Providence. But naively, I never linked these to race or ethnicity. It wasn't until years later, as I was watching *Guess Who's Coming to Dinner?* with Sidney Poitier when I understood the world in 1967. It wasn't until 1967 that the Supreme Court overturned the "miscegenation laws" prohibiting interracial marriage, as unconstitutional.

My parents raised my older sister, Alisa, in Rhode Island for only a short time before moving to southern California, to where the rest of my father's family had now immigrated. It was in California where my older brother, Devon, was born. My dad was able to find work during the day, live in UCLA family housing, and attend night school in the evening, working on his degree in computer science. By this time, he realized that maps were going digital or perhaps more broadly foresaw the future of computing—nearly all of his siblings pursued degrees in math and computer science after coming to the U.S.

Living in Los Angeles in the late 1960s as a Latinx family was a very different experience than what I was raised with by the time my parents had me. When Alisa was enrolled in school, due to her dark complexion, she was automatically placed in a Spanish-speaking classroom. In this age of assimilation, she had been raised as English-speaking only, leaving her crying when she got home from her first day of school. My dad recalls that he would have to carry a copy of his class schedule in his pocket because on his way to night-class he would be routinely pulled over by cops who wouldn't believe he was on his way to class. They lived in California only a few years before returning east.

I was raised in a different America—one nearly 15 years later, on the other side of the United States from my Latinx *familia*. One where my dad finally finished his computer science degree after 18 years of night-school and transferring colleges. One where my mom also went back to college to transform from a 1980s secretary to a school-based speech-language pathologist. One where I was raised in an English-speaking home but took Spanish-classes to be more connected to my Latinx roots. One where I proudly brought my dad's Peruvian trinkets to class for show and tell. One where this Latinx family had now achieved something akin to the American Dream (as long as you don't count the years of school debt, the clash of family backgrounds, and the lack of wealth-building). One where many of my Peruvian aunts and uncles had been successfully educated and employed in the burgeoning computer science industry.

## Finding Mathematics

In an article I co-authored with my mathematician cousin, Marizza Bailey, called "Revealing Luz", I wrote about my journey both towards and away from mathematics.[1] "Revealing Luz" was about my great aunt, Tía Luz, who my dad knew growing up in Peru. She was a math teacher and inspired many school children and family members. The greatest gift she gave me was that my dad grew up knowing that the women in his family had the potential to be incredible mathematicians and educators. As I balanced this line of being Latinx-identified, but separated from most of them on the opposite coast, most of my connection to my mathematical and computational successes as well as my Latinx heritage was through my dad. My dad often spent our time together working on math workbooks while my younger sister, Naomi, was in dance class. My older siblings would show me math flashcards as a baby, however, one could argue an even earlier influence—I am a part of *mi tía* and she is a part of me, and math is a part of both of us. This is a concept Rochelle Gutiérrez brings back to us from Mayan philosophy called *In Lak'ech* in which we see ourselves in others.[2] The version Dr. Gutiérrez introduces in "Living Mathematx" is more an acknowledgement of a broader community interconnectedness and perhaps less literal than what I mean. No matter the reason, my identity as a mathematician is thoroughly wrapped up in my identity as a Peruvian-American.

My dad, ever the influence on my mathematical opportunities, chose a town to move to when he relocated to Boston that had an excellent math program. I captained a math team at a public school with a multi-year winning record in Massachusetts and New England. I was also part of the American Regions Mathematics League Massachusetts A team that won in 1998. I would guess others considered me very successful in mathematics, but I always thought of myself as a modest performer among the best. I clearly loved calculating, but had no interest in studying proof-based mathematics in college. All of my siblings solved this by pursuing degrees in civil engineering, but I wanted to go save the oceans and the rainforest. I did not understand what math could offer me other than a job teaching (which I have to say I loved doing, as by this point I had been tutoring math for years. But I did not love this as much as saving the planet!). My rebellious choice as a headstrong teenager was to reject math, and also reject the thought of applying to any college without a zoology program. My second rebellion was to reject the idea of an Ivy League education—in part because I was turned off by the pomp and circumstance, but also because I did not understand that the price tag was just a suggestion. I was worried about the cost and didn't want my dad to spend the money. At that time, I didn't realize places like MIT have substantial money for financial aid.

---

[1] Revealing Luz: Illuminating Our Identities Through Duoethnography, Eaton, Carrie Diaz and Bailey, Luz Marizza, *Journal of Humanistic Mathematics*, 8, 2,(2018).

[2] Living Mathematx: Towards a Vision for the Future, Gutiérrez, Rochelle, *Philosophy of Mathematics Education Journal*, 32 (2017).

I found my way back to math because I found my way into a community of mathematicians. In retrospect, I had always been looking for my community. Throughout all of my schooling, I felt I never quite fit in and had longed to be in college. When the University of Maine offered me a full scholarship, I found that community and underwent something akin to a metamorphosis. I became involved with various "diversity" groups and found the Latinx and Spanish clubs welcoming of my white face and school-learned Spanish. I credit UMaine, under Dr. Angel Laredo's leadership as Dean of Students, with the intentional development and support of these communities—from its new inclusivity programming to its student leadership programs. I will always remember at our first Latinx festival and dance at UMaine, Dean Laredo gave an opening talk and said that it was *las ganas* (the desire), that drove us to be the best. I also found community in the mathematics department, in the math club, and in my small math classes—unlike my large biology lectures. Within these communities, I found my voice as a leader, and then I found my way back to mathematics. For many, finding math is the difficult part, but for me, it was finding my way back to math since it was a part of me from the beginning, or as we said in "Revealing Luz," finding something that was "always-already-there."

## Balancing Identities

During the summers in college, I worked for a temp agency in Boston. I was placed at MIT in Cambridge and for two summers, worked in the basement of Building 11 as a copy assistant. I had wonderful and funny coworkers and enjoyed it immensely. We were responsible for all of the copying and computer rental time for the math department, and I found myself longingly flipping through past Putnam exams and math course textbooks. There is an odd sort of message looking back on this now—that the Latina in Building 11 was only good enough to make the photocopies. My dad, perhaps seeing this longing and realizing that I was leaning to switch to mathematics, called the MIT admissions office and made an interview appointment for transfer. What my dad did not see was the ragged students copying their thesis who hadn't showered in many days, the campus reports of suicide, and the fact that I felt more at home among the copy staff than the faculty. I also felt more at home among my newly found University of Maine community and family, and I did not want to leave them. Ultimately worried about how I was going to afford tuition and in the midst of developing a community at UMaine, I declined MIT's offer to transfer.

At the beginning of my third year, a few changes began to solidify my new path. I was invited to join a research group and graduate topology course by a professor, Dr. Bob Franzosa. I started dating my best friend, Scott Eaton, and by my senior year, we married. The math department hired a mathematical biologist, Dr. Sharon Crook, who combined both my interests into a single discipline. Finally, the department introduced an interdisciplinary master's in mathematics. Dr. Crook offered me the opportunity to stay as a graduate research assistant in computational neuroscience, which conveniently solved my new two-body problem. It also gave me the opportunity to  move in with my grandparents to help as they aged, despite the long commute. Sadly, in the first semester of my master's program, my grandfather passed away, but we were grateful to be there for my grandmother.

While at UMaine, I continued to take biology courses and became particularly interested in a specific area of mathematical biology called evolutionary theory. At the time, my cousin Marizza was a PhD student in Galois theory. This gave me some confidence to pursue a PhD in order to stay in Maine as a full-time faculty member. This was a difficult leap because, in order to get a PhD in mathematics, I would have to leave Maine. Fortunately, my husband and I were accepted into all of our graduate programs, with generous assistantships and fellowships. We picked the University of Tennessee because it was relatively close, affordable, and offered the same level of interdisciplinarity I currently enjoyed while training in a rather unique area of study.

Balancing family, identity, and a PhD in math was not an easy road. As I once described it, it continued to be like the TV show *Ninja Warrior*—trying to navigate both the planned obstacles while dodging all the surprise ones. I had no idea really about how I was supposed to study for my first prelim at the end of the summer. Even though I excelled in class, I felt like I barely passed the prelim. The final exam for my second prelim class was held on the[3] "Day without Immigrants." I was distracted the whole time thinking about how I felt I could not boycott my exam that day in solidarity. What gave me joy in these hard times was my work mentoring graduate students. I had begun calling for a teaching assistant support program after completing my exams. My proposal to the department was approved, and the next fall I began co-directing the math department's first TA teaching development and mentoring program.

Two weeks before my orals to advance to candidacy and eight and a half months pregnant, my car was hit broadside by a tractor-trailer. I had to be induced and had my son, Gabriel, on July 25th. Then the childcare facility I had reserved space in closed without notice, and I could only find care two days a week. When my son was two and a half weeks old, I was back at the university teaching the TA development class. I called my parents to come help until daycare could start and a fellow graduate student, Erin Bodine, would watch Gabriel while I was in class. When he was not quite two months old, I passed my oral exams, wading through the fog of sleepless nights as a new mom. I was nursing him when I turned down the opportunity to attend an evening awards banquet—where, I later found out, I had been awarded the University's Chancellor Award for graduate teaching.

Nearly five years into my PhD, my husband received a job offer for an instructor position back in Maine. By this time my parents had divorced, remarried, and retired to Maine, and my husband's family also lived in Maine. Much to the surprise of my advisor, I decided to take a leave from my PhD ABD (all but dissertation) to be closer to family support. We moved back in with my grandmother, who by then was no longer grieving, but needed more assistance at home. I nearly quit my entire PhD as I realized I was pregnant with my daughter, Yudani. It had always been good enough to stay here with my family and teach at a local college. However, part of this dream scattered when I realized that the adjunct pay I was earning and my husband's instructor position were not going to be enough to make ends meet. I also felt a sense of regret at the thought of never finishing the last chapter of my dissertation. So, with some babysitting provided by my retired parents, I finished as much as I could on my dissertation independently before

---

[3] On May 1st 2006 various protests and strikes occurred around the United States to demonstrate the importance of immigration to the United States.

Photo used with permission from Goodrich Photography.

Diaz siblings with *papá* in the middle. L to: Devon, Naomi, Hugo, Alisa, and Carrie.

my daughter was born, and went on the job market for a full-time position. As a math biologist in the 2000s, I was lucky to have many interviews in Maine, and I accepted an Assistant Professor position at a small environmental college, Unity College—teaching math to save the environment. It was that position, contingent on finishing my PhD before my first contract review—and the encouragement of my peers—that gave me the last push I needed to finish off the final chapter, which in retrospect was my most independent and intellectually creative work.

Moving home meant my family had more support, and it turned out to be fortuitous timing. My dad was diagnosed with a muscle degeneration disease. My mom was in a severe motorcycle accident just a few weeks before my final dissertation defense and nearly died. She was air-lifted to the hospital I lived only a few miles from, and I was there to support her and her husband as the decision was made to amputate her leg. I went to a conference only a week later, having organized sessions. I have a vivid memory of my advisor asking me if I was ready for my defense, and trying to tell him about my mom while crying on the escalator. My house in those times was a host for all of my family who came to visit my mom in the hospital. In the end, I passed all those exams and all the trials and defenses. It would be years later that I would find out I was the first Latina to graduate with a mathematics PhD from the University of Tennessee. It was an accomplishment of both grit and luck, but not particularly graceful.

Sometimes I feel conflict about the decisions I made on my path. I frequently rejected "superior" choices of institutional name to stay closer to my values in both family and mathematics. However, I do not think I sacrificed any education—maybe connections and access, now that I see the view from Bates, the elite small private college where I currently teach. Perhaps if I did not center family, I would not have faced the roadblocks of being a mother in academia or burdened a significant care-taking role for my parents and grandparents when they needed me. But I know also that if I had made different choices, I would not have been the same person, embraced by my communities, and given the opportunity to lead. I cannot imagine who I would have become had I not made my decisions based on the values that I held dear. I have found this true also for the students I have taught along the way, which have always held the same hidden potential, no matter where they have attended school. My advice for those on the path of self-discovery is to be true to yourself. My advice for everyone else is to look beyond the labels of worthiness that mathematicians use to judge others.

## Interdisciplinarity Means Working Together

Although I describe my upbringing as fundamentally different than my siblings, I was also largely naive and shielded. I still had struggles navigating the differences between my mother's and father's side. We interacted a lot with the Latinx community through my father's church, as my father would often help with translation needs. I have a vivid memory of having a newly arrived refugee come to the house with his daughter. I played with the daughter, while my dad helped the father. Later I asked him about the father because I had heard him crying with my dad. His wife had been shot at the border. Years later, in my suburban English class, I would write about this experience. My teacher scoffed that I most certainly had to be exaggerating. These many differences between worlds also played a huge role in my parents' divorce. My dad could never shake the feeling that he was always trying to prove he was more than the Latino boy that got my teenage mother pregnant.

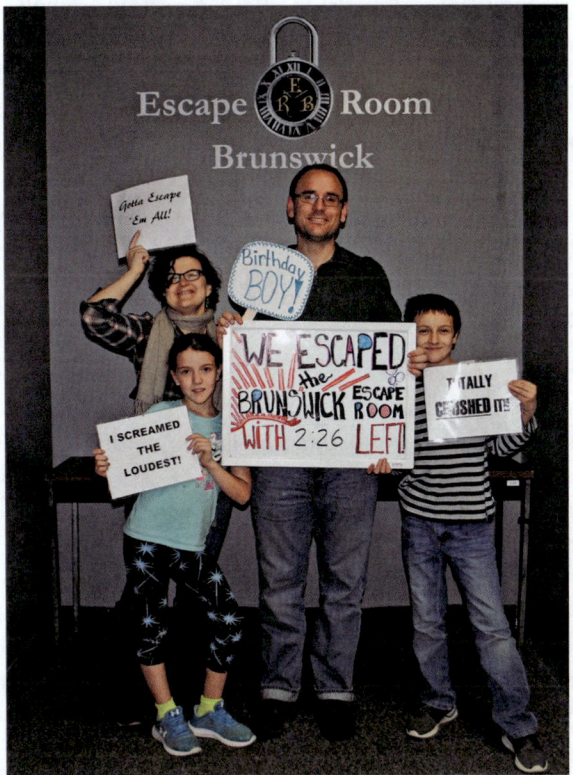

Carrie, Scott, Yudani, and Gabriel in 2018.

There were many worlds my family upbringing straddled—Catholic and Protestant, conservative and liberal, white and Brown, English and Spanish. I only added to these diverse experiences when I moved between mathematics and biology and became a northerner in Tennessee. In addition, I had to navigate these multiple worlds as a mother, as an underrepresented minority in STEM, as a woman, and as a queer individual. These many intersections, worlds, and bridges, created a confluence of experiences by which I became a leader as a boundary spanner. I was used to translating experiences between worlds. I understood the values and axioms that made these worlds operate, and I was consistently looking for common ground between frameworks and ways of seeing the worlds. I was used to thinking about language as reflecting culture and history. Other researchers in Chicana studies and education call this *Nepantla*—the space between these spaces which allows us to think differently and open new spaces of being.[4] Therefore, interdisciplinary research for me is an outlet by which I can use the sum of my knowledge and experience, but also I have found my type of boundary spanning is needed to move research and education forward.

Now I am almost used to moving between disciplines and moving between or embracing multiple identities. This perspective has been particularly important as I have started to work in building an interdisciplinary and inclusive STEM education. How do we talk to biology students about mathematical concepts? What does math feel like to others when it is spoken without first understanding their own cultures and conceptualizations of mathematics? It turns out that biology students already know a lot about math and modeling, they just understand it differently—the engineering examples of calculus speak a different language than what biology students understand. Community co-evolution has also become an important and reoccurring theme. My research in evolutionary theory led me to study plant-pollinator networks—species that work together to build diverse and resilient communities. That final chapter of my dissertation that was almost never written because I left Tennessee ABD was about how mutualistic communities co-evolve together, the structure of those networks, and why they are both diverse and resilient.[5] I want the STEM disciplines to build a diverse, resilient community together.

## Networks for Change

Re-conceptualizing my research strengths as synthesizing frameworks and as about cultivating diverse, resilient, co-evolving community networks has given a coherency to my research program that others' used to label "too-broad" or "scattered." In retrospect, I always naturally gravitated to forming community support networks in my early leadership days at UMaine and with the TA support program at UT even when that work seemed separate from my research.

As soon as the ink was dry on my PhD, I wrote my first grant, to support a network called QUBES (Quantitative Undergraduate Biology Education and Synthesis). We had

---

[4] Embracing Nepantla: Rethinking "knowledge" and its use in mathematics teaching, Gutiérrez, Rochelle, *REDIMAT*, 1, [1], (2012). and *Making Faces, Making Soul/Haciendo Caras: Creative and Critical Perspectives by Feminists of Color*, Aunt Lute Books, (1990).

[5] Diaz Eaton, C. E. (2013). Modeling the Genetic Consequences of Mutualism on Communities (PhD dissertation).

a dream that we would help others find high-quality curriculum materials in math and biology by collaborating broadly with professional societies across disciplinary boundaries and by providing a robust cyberinfrastructure.[6,7] Seven years and about $4 million later, we have moved far beyond our original dreams. I have also been involved in a number of other networks, all with the goal of bringing groups together to collaborate, and more recently with an emphasis on equity and social justice.

I have also been avidly involved with institutional governance and professional societies, which I enjoy. I am not sure how I started, but I have always tried to be a reliable and hard worker and say no when I cannot deliver, and so my name seems to continue to be recommended. I have served as Chair of the Education Subgroup of the Society for Mathematical Biology. I also served on the Board of BioSIGMAA, the Special Interest Group of the Mathematical Association of America (MAA) devoted to undergraduate education in mathematical and computational biology. I am currently the Chair for the MAA's Committee on Minority Participation in Mathematics. I have served on the Editorial Board of *Letters in Mathematics* and currently serve on the Editorial Boards of *CourseSource* and *Problems, Resources, and Issues in Mathematics Undergraduate Studies*. In 2020, I was awarded the Society for Mathematical Biology's John Jungck Prize for Excellence in Education.

## Concluding Remarks

In all of my experiences, my biggest joys have come from raising a wonderful family and from supporting others to be successful on their own terms. I remember being told as a new person in a new institution, that the ideas I had were not going to work. Even though I have seen many ideas rise and fail, I myself have seeded improbable ideas that radically succeeded. Instead of crushing potential, my goal is to always bring out the strengths of the people I work with. The next generation of mathematicians and biologists seems to be more insightful and creative than I have been, and I actively work against the institutionalized ideas that older means wiser, that collaboration means a lack of self-sufficiency, and that niceness fixes systemic inequality. There is still work to do, but I believe we can do it together.

## Advice

I conclude with some advice that I have learned through experience in my life and career. I share it with the hope that it helps readers who are exploring mathematics as a career as well as those who have already found their way to mathematics and have embraced it.

- I think math is a part of us and we are a part of math, but some of us find ourselves on a lifelong entangled path with it. Follow both your heart and your brain. My brain made me pick math over biology as a primary discipline, but my heart brought me to the work I do today.

---

[6] Quantitative Undergraduate Biology Education and Synthesis (QUBES): The Power of Biology × Math × Community (qubeshub.org).

[7] Donovan, Sam, et al. "QUBES: a community focused on supporting teaching and learning in quantitative biology." *Letters in Biomathematics* 2.1 (2015): 46–55.

- Find your support system at school and outside of school, and use it. Figure out what keeps your brain and your heart motivated and use that to keep you going through challenges.

- My advice for those on the path of self-discovery is to be true to yourself. My advice for everyone else is to look beyond the labels of worthiness that mathematicians use to judge others.

- Actively work against the institutionalized ideas that older means wiser, that collaboration means a lack of self-sufficiency, and that niceness fixes systemic inequality. It is easy to be lulled into assimilation, so reflect often.

- Working together, we can build diverse and resilient communities.

# 11

# Dr. Alexander Díaz-López

## When It All Started

It's Saturday evening and the sun is about to disappear from the horizon. I hear the sound of dominoes, clashing with each other. A table is set, my sister and cousins are shuffling the dominoes and we all get ready to play. Sometimes for hours. When (if!) we got tired, we would switch to Monopoly or card games. Regardless of the game, there was one constant: I always enjoyed counting the game points/"money" at the end. In a sense, this is where it all started for me. These are the earliest memories I have about numbers and mathematics.

Dr. Alexander Díaz-López

Illustration created by Ana Valle.

Photo courtesy of Alexander Díaz-López.

The domino playing crew, some years before we actually started playing.

My parents, however, tell a different beginning. They share stories about how I could not speak until after I was two years old but then quickly learned how to speak, read, and count in a very short time; well before starting kindergarten. They tell stories about me extrapolating that if $1 + 1$ is 2 then $10 + 10$ is 20 and $100 + 100$ is 200 at a very early age. Some years later, when I finished second grade, a teacher suggested I skip third grade and jump directly to fourth grade as I had the mathematical skills for it. My parents agreed and I skipped third grade. Unfortunately, I have no recollection of any of these events.

## School and Early College

**El Triángulo de las Matemáticas**. Almost all of my childhood memories start in seventh grade. They are mostly about how I would spend my evenings playing outside with my friends and cousins as, at the time, my parents did not have the resources for me to attend after-school clubs or camps. The memories that do involve math have to do with when we played board games. As a matter of fact, throughout most of middle school and high school, I enjoyed mathematics but never really felt passionate about it. I had the same teacher from eighth to eleventh grade and while I was doing well, the math discussed in class was very mechanical and I never felt engaged or challenged in these classes.

Photo courtesy of Alexander Díaz-López.

The Math Triangle logo in our club shirt, which I still keep.

It all changed when I entered an after-school math club *El Triángulo de las Matemáticas*[1] during my late years in high school. It was led by Mr. Jorge Haddock, who was also my math teacher during my high school senior year. We met every week to work on cool math problems. The thrill I felt after struggling and then solving what at the time felt like very hard problems was an indication that I wanted to keep engaging with mathematics. I felt like I would enjoy doing what Mr. Haddock was doing (and I did not know any mathematicians other than my school teachers), so I decided to apply to college as a math major with the goal of becoming a high school teacher.

**Remembering mom and dad's sacrifices**. My transition to college, in 2007, required a big adjustment. I had just turned 17 when I moved out of my mom's house to live in a tiny college apartment with my best friend in the western part of Puerto Rico. We were both attending the University of Puerto Rico at Mayagüez, a state university known for their engineering programs. As happens to many college freshmen, for the first time I had to take full care of myself (e.g., having to cook, clean, manage finances, set my own schedule, do my coursework, etc). It was then that I started to understand what a big sacrifice it must have been for my parents to take care of me and my sister and provide us with the best possible chance at a college education.

---

[1] The Math Triangle

For instance, in order to pay for our house, Dad entered the military and completed missions in several international destinations. He left home to stay in deserts, camps, and other non-desirable places. Mom held the fort down while Dad was away and then she went back to college in her thirties to complete her bachelor's degree, while working full time and taking care of me and my sister. I realized that if my parents were so strong to go through really difficult times to chase their dreams, I should try my best to do the same.

**Summer research programs**. It was Spring of 2009 and one day, out of the blue, my phone rang. I picked up and a strange but energetic voice said "Hello Alexander, how are you?… Can you read these papers?" Somewhat in shock, I said: "Yes."

Some months before the call, a faculty member told me I should apply to summer Research Experiences for Undergraduates (REU). I applied to a dozen of them, despite the fact that I was underprepared for them. I had only taken an introduction to proofs class and while I worked really hard at the applications, I wrote what I can now confirm were very poor application essays. Not surprisingly, I didn't get accepted to eleven of them. Yet, a twelfth, Dr. Frank Morgan's SMALL REU group needed something special. They needed more than a mathematician. They needed a bilingual mathematician.

In that same year, Frank Morgan had arranged to spend the summer at the University of Granada in Spain and bring his REU group with him. I have never asked him, but my guess is that the idea of having at least one bilingual student was too tempting for him to ignore. So, he looked at my application and called me to talk about the program. When he asked me if I could read some papers and tell him what I thought, I did what I needed to do. I spent hours reading the papers, although I could barely understand what was written in there. A week later we spoke and he officially accepted me into the program.

June came, I packed my stuff and flew internationally for the first time in my life. Once we arrived in Granada, we settled at the *Carmen de la Victoria* university residence, overlooking the imposing Alhambra Castle. At the time, I found it all very impressive. Then, reality struck. During the first couple of research group meetings, the other three participants in the program (all coming from prestigious institutions) were talking about densities, manifolds, and isoperimetric curves. I could not understand much of it. By the third day, I felt hopeless, so I started packing and decided I was going to head back home.

Before making it official, I spoke with Frank and told him I was lost. We sat down and discussed the background needed for me to work on a particular case of the research problem we were working on. More than anything, our

Giving my first-ever research talk in Granada.

Photo courtesy of Alexander Díaz-López.

conversation gave me hope that I could at least attempt to work on a problem. I spent the next four weeks working on it and the problem became part of my first math publication. From that point on, Frank became a big advocate for me and later recruited me to join the *Notices of the AMS* editorial board when he was Editor-in-Chief.

The following summer, I applied and got accepted to the Mathematical Sciences Research Institute Undergraduate Program (MSRI-UP) in Berkeley, CA. This time I was better prepared to work on a research project and had a good experience throughout the program. It was uplifting to do research mathematics in such an amazing place and with an excellent group of peers. To top it off, my research advisor was Dr. Edray Goins, who became a mentor and role model for me.

## Graduate School: Learning from the Hard Times

Finishing my undergraduate degree was my second biggest academic accomplishment at the time (getting the SMALL REU problem solved was the first). I feel blessed to have experienced many other positive moments in my life that fill me with joy every time I think and reflect back on them: getting married, getting my PhD, co-founding Lathisms, getting the job I had desired, among others. While these moments have carried and continue to carry me through life, they aren't the moments that made me stronger. Difficult times are the ones that have taught me how to be resilient.

I rarely talk about these difficult times, but was recently reminded of the power of sharing these moments, particularly, for a younger generation that might look at us and see "awesome mathematicians who rarely struggle." So, here are some of my most difficult academic moments.

**First year of grad school was HARD**. It was an early week in January and the beautiful Notre Dame campus was covered with snow. Temperatures had ranged from single digits to below zero for the past week and I was in my on-campus graduate apartment. I had been inside for about 300 straight hours. I was tired, without energy, and frankly a bit depressed. "How did we get here?" I thought.

A year before that, after the two REU experiences, I was sure I wanted to become a math professor. After a long and stressful graduate school application period, I received a handful of offers and decided to enroll in the mathematics PhD program at the University of Notre Dame. "I am only one degree away from my goal of becoming a university professor," I thought at the time.

The first two weeks of the semester were a slap in my face in many ways. I had the privilege to be raised in a place where I was part of the majority; now, at Notre Dame, there were no Black, Latinx, or Native American faculty in the math department. Overall, there weren't many people of color in the department and more generally, in the university. It felt isolating and it was a sudden introduction to the racial disparities and lack of a path for a graduate education that many people of color face in the United States.

A second shocking issue was classwork. I was not sure what type of mathematics I wanted to study and was unsure about my background, so I thought it would be a good idea to take four courses in my first semester. I enrolled in Real Analysis, Complex Analysis, Algebra, and Topology. Very quickly the semester turned into a stress-inducing machine. Reading notes and books, doing weekly homework, studying for exams. I was spending a lot of time (and doing well) in Algebra and Complex Analysis, but was really struggling with Topology and Real Analysis. In an unhealthy pattern, I would spend most hours of every single day of the week doing course work. Despite this, I kept struggling with Real Analysis. The professor's teaching style, which would have him talk for the whole hour while writing very few things on the board, was not working for me. I obtained less than 40% on both the midterm and the final exam. By being one of the three students who "stuck with it" for the semester (out of the initial 12 or so students), I obtained an A–. But my course average, just like my energy and desire to continue learning from the professor, was much lower than 40%.

With all my family and friends back at home, I was alone, stressed, and tired. Qualifying exams were coming in January and temperatures were ranging in the single digits. I decided to quarantine myself in my apartment to study for the qualifying exams for twelve straight days. This is how I ended up inside for about three hundred straight hours, exhausted, and depressed. By the end of it, I took my qualifying exams and felt like I had not only dumped all the stuff I had just studied, but that I also left all my humanity there.

As soon as I turned in the exam, I knew I was out of energy and motivation and needed help. After taking a week-long break, I reached out to my community; I talked to my girlfriend (who is now my wife), advisors, counselors, and colleagues. The consensus was that I needed to make changes; otherwise I would not survive graduate school. I started exercising and playing sports, stopped prioritizing coursework over my own physical and mental health, and reduced the coursework to a manageable load and in areas I was more interested in. Fortunately, I really enjoyed Katrina Barron's Algebra class and so I started taking more courses in the area. Not surprisingly, I ended up getting my PhD in algebra (under the direction of Matthew Dyer).

**Withdrawing from differential geometry**. The lessons I learned during my first-year graduate experience helped me get through the remaining years of my graduate career. For the most part, years two through four were fairly positive, in part, thanks to the advisor and area of research I chose. But there was one more experience worth pointing out. For the first time in my life, I had to acknowledge that I was going to fail a math course. I had a 3.90-ish GPA in high school and college and had never really failed a course. Even graduate Real Analysis, with the low scores I had, I was fairly sure I would pass as every other student was in the same boat. Then I met differential geometry. After the first two weeks of classes, it was clear to me that either I devote all my time and energy to that class (mostly to pick up all the needed background and then to catch up in the class) and relive what I lived through during my first year or I was going to fail the course. I had it clear at the time. There was no way I would go back to the unhealthy habits of year one. Hence, I accepted I was going to fail and decided to drop the class.

**Job search**. Going into my fifth year, there had been one question I was dreading to ask my advisor. "When do you think I will finish the program?" After a year struggling to get myself to ask it (and afraid of the potential answer as I had guaranteed funding for exactly five years) I did ask him. Matthew, quickly asserted: "If it all continues well, you can finish this year."

I felt so relieved. I could finally finish and fulfill my goal of getting a PhD and becoming a professor. Yet, one more difficult process lied ahead—the job search. Looking for jobs in academia is a nine-month process. First, drafting your documents (in August at the latest), then searching open positions, submitting applications (early winter), getting interviewed (January), doing on-campus interviews (February/March) and finally accepting a position. It's a draining process, to say the least.

I was inexperienced in the process and had no training about it. So, I asked my colleagues what they planned to do and many said they would apply to 100+ jobs. I then did the same. It wasn't until a year later that I realized this was a terrible idea, at least for me. How can one really research 100 places, gather information about their positions, the type of department they are, what they are looking for and then tailor every single one of the applications? I don't find it possible.

Not surprisingly, I received 80+ rejections. Well technically less, as many of the places never actually contacted me with a formal rejection. Fortunately, I seemed to have been an exciting candidate to some liberal arts institutions, as I got visiting offers from Swarthmore College, Haverford College, and Williams College, and a tenure-track offer from Hobart and William Smith Colleges. After visiting these places, I decided to decline the tenure-track offer and join the Department of Mathematics and Statistics at Swarthmore College. It was a risky move, but it ended up being one of the best decisions I made. The year I spent at Swarthmore confirmed that being a professor is the perfect job for me and a year later I obtained a tenure-track position at Villanova University. The three years I have spent at Villanova have been professionally and personally fulfilling.

## Community Building

As a Latino who grew up in Puerto Rico, I benefited from a culture in which building and fostering communities was always important. Yet, the somewhat individualized experience I had in graduate school and some solitary experiences at the Joint Math Meetings made me wonder if it would always be like that. Thankfully, it has not.

**Meeting Erik, Darleen, and Pam at USTARS**. The Underrepresented Students in Topology and Algebra Research Symposium (USTARS) was the first conference I attended in which I felt I belonged. Instead of being the single or one of the few people of color in every single event of the conference, the whole conference was centered around us. More

USTARS participants and organizers in 2014. Photo provided by the USTARS 2014 organizing team.

importantly, there was a concerted effort to create a supportive and welcoming atmosphere for the graduate students involved in the conference.

In that place, I started building a community. I attended a talk by Darleen Perez-Lavin, which led me to also meet Darleen's collaborators Pamela E. Harris and Erik Insko and which later resulted in a research collaboration between all of us. Since then, Pam, Erik, and I have worked on numerous research projects. Through the connections I made there, I also met Mohamed Omar, Alicia Prieto-Langarica, Gabriel Sosa, and many other individuals who have been influential in my career.

The biggest lesson I learned is that building community does not happen by chance. It requires planning, purpose, and commitment from the individuals involved. And once you build communities, they can be life-changing. Thus, I have decided to get involved in projects which, in one way or another, are centered in building community.

## Lathisms: Latinxs and Hispanics in the Mathematical Sciences.

During a conference, Pamela E. Harris, Alicia Prieto-Langarica, Gabriel Sosa and I discussed the idea of showcasing the contributions of Hispanic and Latinx mathematicians during Hispanic Heritage Month (HHM) as a starting point for building a stronger and more connected Hispanic and Latinx community in mathematics.

Shortly thereafter Lathisms (Latinxs and Hispanics in the Mathematical Sciences) was born. During the first year, we showcased one mathematician per day during HHM. They are featured in the poster to the right. Since then Lathisms has created a podcast, listserv, and continues to showcase Hispanic and Latinx mathematicians on its website. With support from the American Mathematical Society (AMS) and Mathematical Association of America (MAA), the project has reached thousands of schools and many universities.

Lathisms poster produced by the American Mathematical Society.

Photo courtesy of the American Mathematical Society.

## Villanova: DREAMS Program, Co-MaStER.

One of the nicest things about Villanova is that the faculty in the department have a great rapport with each other. Yet, our connection to math majors is not as robust. To bridge the gap, I co-created two programs: DREAMS (Discovering Resources and Exploring Advanced Mathematics and Statistics) and Co-MaStER (Community of Mathematicians and Statisticians Exploring Research).

DREAMS' goal is to introduce students to intriguing problems in mathematics and statistics that could potentially lead to a research project and to provide graduate school information and advice, along with mathematics career perspectives that demonstrate the value of advanced studies in math. Co-MaStER is a research program where we gather research projects from different faculty in the department, send them to students, collect students' applications and pair interested students with appropriate research projects. Each research group functions independently, but once a month everyone gets together for professional development and research sharing sessions. So far we have received overwhelming positive feedback about them.

**Math SWAGGER**. In Summer 2020, I joined Pamela Harris, Vanessa Rivera Quiñones, Luis Sordo Vieira, Shelby Wilson, Aris Winger, and Michael Young in co-leading the Mathematics Summer Workshop for Achieving Greater Graduate Educational Readiness (Math SWAGGER), a five-week virtual summer program for underrepresented students enrolled in a mathematics/statistics graduate program. We engaged in conversations about topics that affect the life and academics of graduate students. In addition to the fact that we all learned from each other and from our stories, advice, sufferings, and successes, the early results so far point to the creation of a strong community of scholars ready to support each other through their own graduate paths.

## Conclusions and Advice

Through all the programs I have been part of (both as participant and organizer), it has become clear to me that while no individual can single-handedly create a space where all members feel welcomed, heard, supported and given an opportunity to thrive, when we all come together as a community of engaged scholars with a common purpose we can create such events and spaces. When creating these spaces, it is imperative to think about who are the most vulnerable members of our communities and how can we make sure we provide the tools for such members to excel.

A second and final concluding thought, which I shared in a 2020 interview for the Meet a Mathematician series, is about the idea of being successful. In the interview, I was asked to provide some words of wisdom to the mathematics community. My response: "Don't let others define what success is for you." Success can have many different forms in the mathematical community: being a professor at a research university, a liberal arts college, or a teaching college (see Lathisms.org for many examples), teaching at the elementary, middle, or high school level, becoming a journalist or freelance writer (e.g., Evelyn Lamb), leading community centers (e.g., Amanda Serenevy), running nonprofit organizations (e.g., Jeanette Shakalli), working in industry or government, and many others. Choose what is more appealing to you and define success for yourself.

# 12

# Dr. Stephan Ramon Garcia

According to certain metrics, I am a highly successful mathematician. In 2019, I was elected a Fellow of the American Mathematical Society and was awarded the inaugural AMS Dolciani Prize for Excellence in Research (see the photo on page 125). I have been on the editorial boards of well-known journals and written over a hundred papers, along with several books with top publishers. I have an endowed professorship at a top-tier liberal arts college, and I have received multiple NSF grants and six teaching awards. So it would seem that I know what I'm doing. However, I was

Dr. Stephan Ramon Garcia

rather clueless for much of my journey and often muddled through without clear direction. Nothing about my career was inevitable: some favorable circumstances, fortuitous timing, coincidences, and good luck played important roles, along with lots of hard work.

## Early Life

My father's family fled to the United States from Cuba in 1960. They had to quickly adapt to life in Miami. My grandfather worked as a Spanish teacher and my grandmother as a hairdresser. My father studied at Miami-Dade Jr. College and earned an associate degree in electronics.

My mother was born in Hiroshima during WWII and is an atomic bomb survivor. She attended some college in Japan but did not complete her English degree. In 1969, she was visiting her sister, who had married a U.S. marine, in New Jersey. My parents were set up on a blind date and the rest, as they say, is history.

I remember little about life in New Jersey, save that I would eagerly hop up on a wooden box to look out the window to see my father returning on a motorcycle from his job as a technical aide at Bell Labs. He went to night classes and got a bachelor of Science in electrical engineering from Newark College of Engineering. When I was a toddler, my father secured an engineering position in San Jose, California. The city was years away from its emergence as a globally-recognized center for technology. At the time, it was a sizable, but unremarkable, city in the shadow of nearby San Francisco.

Our new neighborhood was lower middle-class, with cars propped up on blocks in driveways and frequent petty crime. The nearby public schools were mediocre, which

Photos courtesy of Stephan Ramon Garcia.

(L) My paternal grandparents, and my father and aunt (Cuba, 1950s). (R) My maternal grandparents (Japan, 1946).

caused my parents to enroll me in a series of private schools. I was an only child and my parents were frugal, so this was just barely affordable. I recall little about my first few schools, save for a vague sense that I did not belong there.

From third to fifth grades, I attended the local public elementary school. My third-grade teacher gave me more advanced math books, and I started learning algebra and geometry. I was placed in a special program for "gifted and talented" students and bussed once a week to another school for enrichment activities.

We moved to a different neighborhood when I was in the middle of fifth grade. Although the public schools there were much more highly rated, it was at first a step back for me. Instead of being viewed as "gifted," as I was in my old neighborhood, I was largely overlooked, even being placed in the lowest English class. Being one of the few Latinx students in the school, I stuck out and was frequently picked on. Fortunately, I moved to middle school after a few months, which was a welcome reset. There I was a good, but unexceptional student in everything but science (I was awarded the prize for best science student at my middle-school graduation).

My high-school career was unremarkable. I was the only Latinx student in most of my classes. I did not know how to learn although I was adept at going through the motions. Although I received a 5 on the AP Calculus BC examination,[1] I did not truly understand calculus. I simply knew how to robotically perform computations.

My parents were immigrants with little knowledge of higher education in the United States, and we did not know how to "play the game" of college admissions. It was natural for us to assume that good grades and test scores were the keys to success. Now that I work at an elite private college, it is clear to me that admissions committees look at

[1] From collegeboard.org: The AP Calculus BC Exam will test your understanding of the mathematical concepts covered in the course units, as well as your ability to determine the proper formulas and procedures to use to solve problems and communicate your work with the correct notations. According to the College Board a 3 is 'qualified,' a 4 'well qualified,' and a 5 'extremely well qualified.'

much more than grades and test scores. Students are often so "well rounded" that they can appear nearly spherical and paradoxically featureless.

I had no curiosity or desire to attend school outside of California, and I applied only to large universities in state. We knew nothing about small liberal arts colleges.[2] I did not get into Stanford, despite excellent grades and test scores. I was simply not "well rounded" in the way that admissions committees valued. I had no student-government experience or athletic achievements; I started no clubs, performed no volunteer work, starred in no plays, and gave no solo concerts.

Fortunately, my grades and SAT scores were high enough for me to gain admission to UC Berkeley with a UC Regents Scholarship. Berkeley was only an hour from home, so I never considered any of the other schools to which I was admitted.

The Garcia family in the 1970s. Seated are my paternal grandparents and I. My parents stand at the upper right. My aunt and her husband stand at left.

## Undergraduate Education

I had no idea what to do at UC Berkeley. There was no internet to speak of: no course reviews or "frequently asked questions" pages were available. I simply signed up for classes that roughly mirrored my high-school schedule. It seemed natural for me to take English, history, math, and science. Nobody told me to do otherwise.

Early on, I got the top score on an exam in a 500-student mathematics class. Although there was no sign that the professor paid any attention to me or my performance, this was my first genuinely positive mathematical experience since elementary school. Brimming with confidence, I received a middling score on the second exam, although I did get an A in the end.

Overall my work ethic as an undergraduate was poor. I spent most of my time playing video games or Dungeons & Dragons, practicing guitar, or playing basketball. I did as little work as possible while still earning a decent grade. Instructors did not notice or care that I skipped class frequently. In a "small" 100-student class, who would notice my absence? Certainly nobody ever reached out to me about it.

By my sophomore year, I was leaning toward being an English major. I did not realize this at the time, but it was my burgeoning competence in mathematics that made me

---

[2] Years later, when my parents moved out of my childhood home, I found recruitment letters from Pomona and Harvey Mudd. None of us knew at the time that I would end up in Claremont.

effective in my English courses. Essays and proofs are not so different: in both cases one makes a logical argument, supported by facts (either quotations from the text or previous results), intended to convince the reader that the thesis is correct. However, I quickly grew disillusioned with the English major. The books that my instructors selected felt increasingly motivated by cutting-edge literary fashion (which I did not care for), and I felt that I was being rewarded for inventing interpretations of texts that, in my heart, I knew that the author did not intend.

In the absence of advising, I continued my old "high-school schedule" although I dropped English courses in favor of history and music. All the while, I was taking physics and mathematics (I eventually realized that what I liked about physics was the mathematics). As I progressed through linear algebra and differential equations, I developed a genuine interest in mathematics. I started reading popular titles like *Infinity and the Mind*; *Gödel, Escher, Bach*; and *The Man Who Knew Infinity*.

My initiation into proof-based, upper-level courses was a shock. I took abstract algebra and real analysis first because these courses were labelled the lowest of the upper-level courses: 104 and 113, respectively. My abstract algebra professor was an unenthusiastic postdoc who lectured out of the book; the 8 AM time slot ensured that I did not attend often. On the other hand, my instructor for real analysis was spirited and engaging. I would lean toward analysis and away from algebra for several years because of these experiences. This experience highlights an important lesson for students: a good professor can make any topic interesting and a bad professor can make any topic uninteresting. Don't judge a topic based upon one course!

Mathematics, in which the grades depended entirely on homework sets and exams, proved to be a good match for my largely nocturnal existence. There were no labs or discussion sections, and professors did not seem to complain so long as my assignments were turned in and I showed up for examinations. In retrospect, perhaps I was turned off by the posturing and preening of some of the other students.

I briefly attended the Putnam[3] seminar, but did not find the atmosphere inviting. The professor in charge had a gruff demeanor and talked mainly to a few international students, all of whom seemed to have had previous math-contest training. I decided that the Putnam was not for me (or perhaps it was decided for me).

Being of mixed race ensured that finding community was difficult. It forced me to be independent in a sink-or-swim fashion. It never occurred to me to want role models who were like me. From what I could tell, there were no people like me. Fortunately, these feelings never hampered me academically. It just seemed natural, indeed obvious, that I would be different from other students and my instructors. This probably helped me navigate a large, impersonal place like UC Berkeley.

Since I had spent so long settling on a major, I needed to double- or triple-up on mathematics courses in order to graduate (it took me five years). I did well in most classes, stood out in a few, and was mediocre in some others. In particular, I earned top marks in a graduate-level analysis class, which proved fortuitous since the professor later served on the graduate admissions committee when I applied.

---

[3] From the Mathematical Association of America website: The William Lowell Putnam Mathematical Competition is the preeminent mathematics competition for undergraduate college students around the world.

Soon my senior (technically fifth) year loomed on the horizon. I wanted to avoid the real world; getting up early in the morning and wearing a tie sounded dreadful. So I applied to graduate school.[4] I had no idea what being a professor was like, nor did I know anything about research in mathematics. I liked mathematics and was intrigued by it, but I hardly knew what being a mathematician meant.

I recall one conversation with a professor. He told me not to apply to top-tier places on the East Coast. He explained that Princeton, Harvard, MIT, and the like were out of reach since I had only taken two graduate-level courses, and I was neither a Putnam Fellow nor an International Mathematical Olympiad medalist. So I applied only to a few schools in California. My performance on the GRE, overall good grades, and the Berkeley name, ensured that I made it to the next stage.

## Graduate Education

I was accepted by every graduate program to which I applied, although that is hardly an accomplishment since I applied only to a handful of schools in California. I had some satisfaction in rejecting Stanford's offer; they had rejected me as an undergraduate. In retrospect, I might have benefited from their smaller program. However, at the time the mathematics PhD program at Berkeley was tied for number one in the nation, so I did not seriously contemplate leaving for slightly lower-ranked Stanford. After all, Berkeley was familiar and Stanford seemed so distant.

There were ten of us assigned to two adjoining offices in the windowless corridors of Evans Hall. Of these, I think only two or three of us completed the program; at least five quit or were kicked out. There were a few other Latinx graduate students in the department, but they all seemed to have been the top students in their countries and many had experience in the International Mathematical Olympiad.

Because I already had a circle of friends in Bay Area, I did not hang out in the math department. Consequently, I did not learn useful tips from other graduate students or from postdocs and professors. Since I did not understand the titles or abstracts, I did not attend colloquia or seminars. I failed to integrate myself into the social side of mathematics. I simply had no idea how mathematicians socialized or learned. The department at Berkeley was large, and it was possible to disappear completely, which I did. Nobody told me what I needed to be doing, and I got lost.

Because I had an NSF Fellowship, I did not need to teach. However, I asked if I could teach one course per semester. It seemed like a good idea to have teaching experience since, I imagined, teaching was an important part of being a professor. I went on to win several teaching awards at UC Berkeley, which opened a few doors.

The transition from taking classes to doing research was left largely unexplained. Since I liked analysis and had just taken complex analysis with Donald Sarason, I asked him to be my advisor. He agreed without hesitation. Because of my lackluster performance in the program and my sparse attendance at department events, I suspect that many other potential advisors would have politely excused themselves.

---

[4] In hindsight, this was a good idea. I graduated in 1997, as the tech industry was booming. Many of my friends went into industry, only to be laid off or have the startups they worked for go belly up with the burst of the dot-com bubble in 2000. It took some time for them to find their feet again, only to get wiped out again by the 2008 crash.

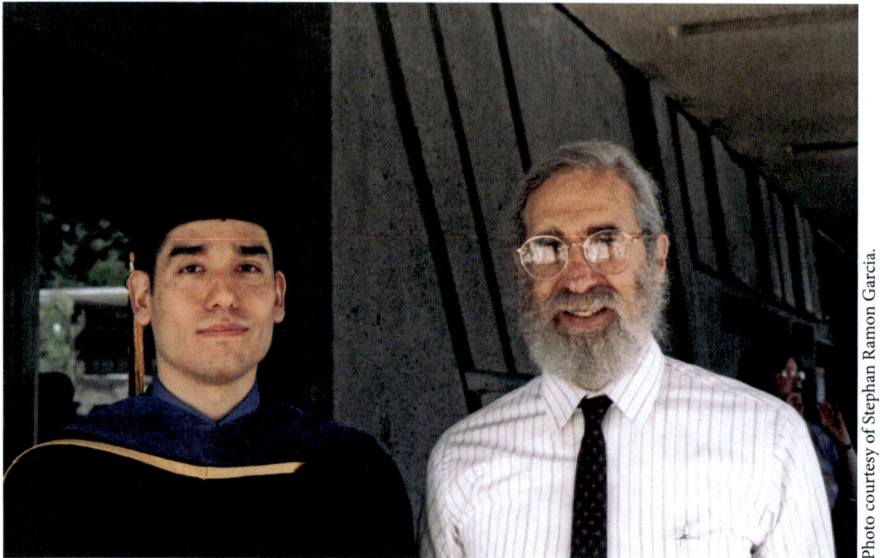

Photo courtesy of Stephan Ramon Garcia.

With my thesis advisor, Donald Sarason, at my graduation from the PhD program (2003).

Sarason, then nearing seventy years old, was kind and patient, but unusually quiet. His advisor, Paul Halmos, said "*[he] is a quiet man; he never uses eight words when seven will do.*" Perhaps a more astute career move would have been to attach myself to an up-and-coming star, swimming in grant money and fresh off an International Congress of Mathematicians (ICM) lecture or major prize. However, the preening roosters and showoffs were attracted to such advisors, and it is not clear that I could have flourished in such a competitive environment. Somehow things worked out for the best.

My qualifying examination committee consisted of Sarason, Michael Christ, and Vera Serganova, along with an engineering professor who admitted sheepishly that he was just there as an outside observer. Although I answered the first few questions well enough, things turned for the worse. Christ is a tall, imposing man with a deep voice, and I felt somewhat intimidated when I fumbled one question. Serganova asked a few algebra questions in a kindly fashion, perhaps taking pity upon me, or possibly just throwing out softballs since algebra was the minor topic on my examination. After a long several minutes in the hallway, I was informed that I had passed, although I certainly felt that I hadn't or shouldn't.

My fifth year of graduate school rolled around, and I had accomplished relatively little. Without formal coursework or well-defined goals, I had spent most of my time in graduate school on non-academic endeavors, although I had apparently done just enough to convince the department that I was worth keeping around. Probably I could have used a swift kick in the rear or stern words from some authority figure.

I had no idea how to be a mathematician, no idea how to do research. I had never been to a conference, nor had I met key players in my field. By some stroke of luck or inspiration, I managed to put together a decent thesis. Although my dissertation solved a problem from an old *Bulletin of the AMS* article, it was written so abstrusely and tersely that it gained little traction. I briefly met one of the authors of the *Bulletin* article at my advisor's seventieth-birthday conference. However, I could not succinctly express my ideas: I was a poor mathematical communicator. The professor appeared impatient with my rambling, and I received a cool and critical response. Clearly, I had no idea how to give an elevator pitch.

Sheldon Axler probably saved my professional career. As a clueless graduate student just about to hit the academic job market, I needed letters of recommendation. But I didn't know anyone! Sarason reached out to his former student, who was, fortunately, willing to meet with me. I traveled to San Francisco State University, where Axler had relocated as department chair after a distinguished career at Michigan State. Fortunately, he entertained my rambling and incoherent explanations long enough to see that there was something worthwhile behind the nonsense. He wrote a letter for me which, I can only assume, was a decent one.

I had survived graduate school, if only barely. A last-minute thesis breakthrough and my advisor's connections had saved the day. What next?

## After Graduate School

I was fortunate enough to obtain a postdoctoral position at UC Santa Barbara. My girlfriend, Gizem Karaali, completed her PhD at UC Berkeley the following year and also secured a position at UCSB. We married soon after.

My mentor at UCSB was Mihai Putinar. Although the graduate students dreaded him as the "demanding Eastern European analysis professor," I found him to be a prolific mathematician with a broad perspective. He gave timely advice about mathematical politics, grant writing, and all aspects of the profession. We wrote several influential papers during those years and I really came into my own as a mathematician. I probably would not have succeeded without Mihai's guidance.

With my mother (left), paternal grandparents (middle), and Gizem (right) at my graduation from the PhD program (2003).

Photo courtesy of Stephan Ramon Garcia.

Our daughter with her great grandparents in Miami (2010).

Although I spent most of my time on research, I won another two teaching awards at UCSB.[5] Moreover, I turned my bad habits around and became a workaholic: it was the only way to survive the publish-or-perish academic job market. Now that I knew what I was doing, there was a lot of catching up to do! Moreover, I felt that I had to work twice as hard for half the recognition: I was not in a "hot area" at the cutting edge of fashion, and I had to struggle against the constant perception that I was not a "real" mathematician and just there for window dressing.

The economy was still humming in 2005, with the Great Recession several years away. Gizem and I were fortunate that hundreds of tenure-track positions were advertised that year; we each applied to over one hundred. It strikes me even today how one's career opportunities depend upon the vagaries of fate. We had several pairs of job offers, along with multiple single offers. We were in a strong position with plenty of bargaining power. Would we be mathematicians today if we had applied in 2008? Perhaps not.

## Looking Forward; Some Advice

As a Cuban-Japanese person from New Jersey, my life story is hardly universal. Nevertheless, I think that we can still identify some counterproductive behaviors, unfortunate incidents, and repeated mistakes from which we can extrapolate some useful general recommendations.

First of all, don't let other people limit your options. Don't let people tell you what you are capable of, set your limits, or deny you opportunities. You can be your own best advocate: if you don't believe in yourself, others are unlikely to step up and go to bat for you.

Second, get your head out of the sand. Meet people and socialize: mathematics is a

---

[5] Upon my arrival at UCSB, one of the senior faculty members advised me "don't spend too much time on teaching. You are here to do research." Because I had won two teaching awards at UC Berkeley, there was apparently some fear that I was not serious about research.

social endeavor. Don't be afraid to ask questions; if you don't ask, you won't find out the answer. Learn from other people and network, network, network! There are lots of things that "everyone knows," but nobody tells you. If you isolate yourself, then you won't learn the ropes and you'll get left behind.

Lastly, focus and work hard. The first stages of one's mathematical career are difficult and stressful; for someone swimming upstream doubly so. Whatever you do, put in 111% (since you'll have to outwork those 110% folks). Sometimes you will have to do more work for less recognition. You'll eventually earn your place at the big table and then you can pay it forward and lift up the people behind you.

Although I've done a bunch of things over the years, I believe that my biggest impact has been in the classroom. Students look to you as a role model and mentor, but more importantly they look to you as the one-stop shop for part-time jobs in the department, research opportunities, graduate school advice, letters of recommendation, and emotional support. You are the one who needs to tell them the things that "everyone knows." You are the one who needs to ensure that they don't make the same mistakes that you did. Once you figure something out about how the world works, make sure your students know!

I mentioned times when teachers thought lower of me because of my background or when professors, perhaps inadvertently, dissuaded me from pursuing opportunities. I still occasionally find myself in settings that are uninviting, in which people view me as necessary decoration, a nod to diversity. You just have to prove people wrong. Once you get to the big table, don't be afraid to stand up for yourself or voice your opinions. Most importantly, find like-minded individuals and mentors. Others have been there before you, so make sure to draw upon their collective wisdom!

Although my "success" was not pre-ordained, I did have some lucky breaks. I was fortunate to have parents who valued education and a stable home environment. Both of my

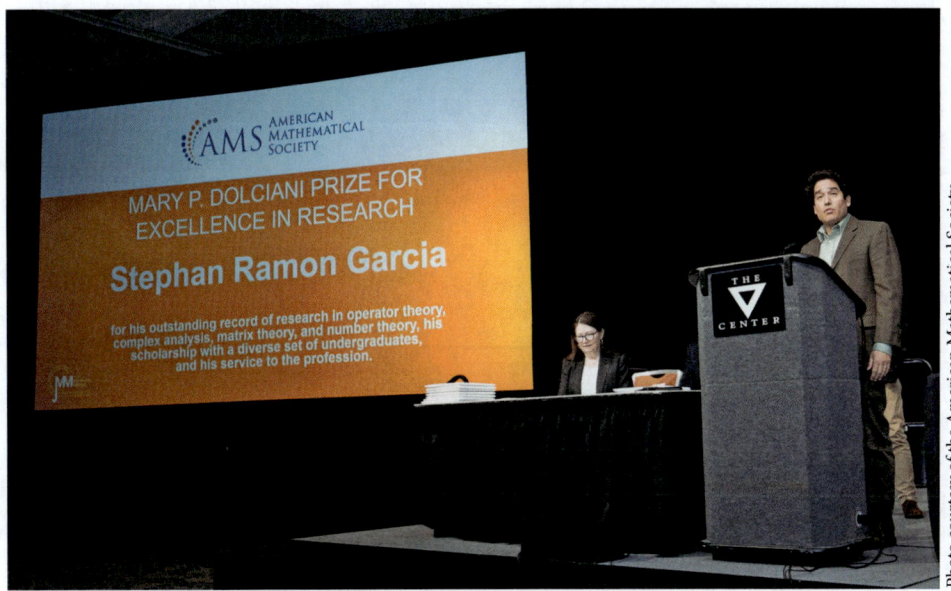

Receiving the inaugural AMS Dolciani Research Prize at the 2019 JMM.

parents overcame poverty and suffering; I benefited from the opportunities they struggled to give me. The Berkeley stamps on my diplomas carried significant weight at crucial moments. My advisor's connections gave me a last-minute reprieve when I needed another letter of reference. At UCSB, I found exactly the right mentor at the right time. Moreover, the global economy cooperated with our job searches.

Even though I now have many of the trappings of "success," my journey was neither inevitable nor without difficulty. There were moments of indecision, self-doubt, and discouragement. I hope that students reading this will realize that even those who seem to know what they are doing may have once been lost themselves.

# 13

# Dr. Ralph R. Gomez

*I would like to dedicate this article to the memory of my mother Sally Gomez,
my father John Gomez, and my brother Johnny Gomez.*

## Early Years

**Lemoore**. About forty miles south of Fresno,
California, in the Central Valley is a small town
called Lemoore. This town is but one of many little
towns that comprise the majestic farm-field tapes-
try of the valley. My grandparents, parents, aunts,
and uncles all worked portions of these vast fields
for many years as field workers picking plums,
grapes, apricots, and peaches.

Dr. Ralph R. Gomez

Illustration created by Ana Valle.

I lived at the dead end of a street which spanned
six blocks. My house was uniquely nestled between
a corn field and the cheese manufacturing factory,
Leprino Foods. Because of the house's close proximity to the factory, the constant noise
and agitation of diesel trucks shuffling about could be heard all hours of the day. Though
my street was only six blocks long, it was regarded as one of the bad parts of Lemoore,
primarily because of the gang activity in the area.

In my family, I have two brothers, Johnny and Eric, and a sister Gloria (Johnny passed
away in 2015). I am the youngest of the family. After being in the Navy for three years
my father, John Gomez, worked in a mill called Continental Grain and kept that job for
forty-two years until his retirement. My mother, Sally, stayed home and ran the entire
household. When she was growing up, she had to drop out of high school to help take
care of her younger brothers and sisters. Thus, she had a plethora of experience in helping
others and maintaining a household.

My oldest brother Johnny and I were seventeen years apart in age. By the time I was
born, he had already moved out of the house, so I saw very little of him growing up.
However, the times in which I did see him usually were not the best of circumstances.

**My brother Johnny**. Without a doubt, Johnny was one of the most influential forces
that launched me on my trajectory to eventually become a mathematician. This attribution
of influence is not because he was a pristine role model with infinite pearls of wisdom and

Johnny Gomez, my oldest brother.

prophetic advice. On the contrary, his life became a living example of the kind of life I did not want to lead. Exactly the opposite.

Many of my early memories of Johnny revolve around going to the Kings County Jail with my mother and sister to visit him. He was in and out of jail for much of his life. I can vividly recall how one of my first visits to see him deeply and profoundly impacted me. When I was around eight or nine years old, my mom, sister and I went to visit Johnny in jail. As we entered the visitation room, we saw Johnny behind the glass partition, seated and waiting for us to take our seats. I remember looking at my mother's face as she looked at Johnny. I could instantly read the complete heartbreak and the immense disappointment she had for him. It was during this time that I decided to follow a path which was in the exact opposite of my brother's path if for no other reason than to spare my mother any more heartache and pain. Thus, I decided to embrace education and stay out of trouble.

**Skateboarding**. Admittedly, it was sometimes challenging to remain on my self-selected path of embracing education and staying out of trouble. On my street was a gang called XIV (the fourteeners), which seemed to be based on one of the more notorious gangs in

An "ollie" is a skateboard move in which the skateboarder jumps in the air with the skateboard, unaided by the hands. Here I am doing an ollie over a traffic barricade in Hanford, California.

Los Angeles. Around my sophomore year of high school, I seriously considered joining this gang on my street. Some of my friends had joined, and I became attracted to the gang's stature and cohesion. But fortunately for me there was a very welcomed diversion and that diversion was skateboarding. Though I had ridden a skateboard since the sixth grade, I began to take the sport very seriously in high school. During the school day, I concentrated hard on my academics. After school, I practiced skateboarding with equal intensity. Skateboarding became my primary outlet from academics, an avenue to forget all of the daily worries and focus on improving my skateboarding skills.

From skateboarding, I sustained a myriad of collisions with the concrete; numerous sprained ankles, two broken bones, a sprained back, and many scrapes and head gashes. Hopefully I learned something from all of these injuries! Skateboarding gave me the necessary structure and the motivation to constantly improve at something. It allowed me to experience the triumph and satisfaction of learning a trick after countless failed attempts. It taught me to cultivate a deep sense of concentration and focus on something of interest—a quality that was crucial once I found my passion in mathematics.

# Higher Education

**Undergraduate experience**. In my household, the idea of going to college was never discussed or mentioned. No one in my family attended college (although my brother Johnny was always fond of saying that he was a graduate of Penitentiary State!). College was an idea I only saw on the television. In high school, I took college preparatory classes, but there was never any intention of attending college. Taking such courses simply reinforced the idea that I was embracing education and attempting to extract the most I could from my small-town public high school. I took the usual advanced courses in English, biology, chemistry, and math. The highest math course I took in high school was precalculus. I had no idea what I was doing, nor did I have any interest in the subject.

In my senior year of high school, I decided to enroll as a full-time student at the local community college, West Hills Community College, after graduation. I had no other plans and going to West Hills allowed me to stay close to my friends who were also enthusiastic about skateboarding. During the time I started college the campus was quite small, comprised of a main building and a few portable buildings. I actually preferred to attend College of Sequoias which was another community college further away from home. It had more course offerings, but was over thirty minutes away by car. My father said there was no way the family could afford all that gas as well as the prolonged use of the family car. Thus, I enrolled at West Hills Community College, which allowed me to walk to class and back home. These walks were actually quite useful because I used to lecture aloud to myself on scientific topics while walking home. I did a lot of homework at the community college library since it was difficult to concentrate at home.

In my first year at West Hills Community College, the placement exams recommended I take trigonometry. My first math class at the college level was trigonometry! This was exactly the right starting course for me. It was during this time that I really started to take mathematics and science very seriously. I realized that doing a whole bunch of problems thoroughly and clearly greatly helped me solidify my understanding of topics. I also found that explaining the idea aloud to myself really helped me in absorbing ideas.

My time at West Hills was an exceedingly pleasant one. I had a few inspiring professors at the college. For example, my chemistry professor, Dr. Robert Holmes, was one of the few instructors with a PhD at the college and was a very encouraging person to me. He was eloquent, highly scientific, and very serious. After earning top marks in his chemistry course he gifted me a chemistry book as a form of encouragement to continue. This inspired me to major in biochemistry once I transferred to a four-year university.

I also took a history course that was highly influential in my future interests. For the final paper in the course, I had to write about an event that changed world history. I chose the topic of the development and use of the atomic bomb. During my research for the paper, I began learning about some of the scientists associated with the bomb. This in turn led me to read about some of the key contributions Albert Einstein made to physics. Soon after, I became extremely interested in the many wonderful ideas in physics. It was during this time that I began to realize the full power of mathematical thinking—its ability to describe nature. It motivated me to want to learn about general relativity and quantum mechanics. I was completely overcome by the magical beauty of how equations could so simply describe physical processes. I was particularly struck by just how geometric the universe actually is. But if I was going to delve deeper into the physics, I had to learn much more mathematics.

After two years at West Hills Community College, I transferred to the University of California at Santa Cruz. Initially, I was a biochemistry major. The transition from community college to the university was incredibly difficult for me. In fact, after the first semester, I seriously considered dropping out of college. I could not keep up with the fast-paced environment and much higher demands of homework. Most of all, I grew very disheartened by laboratory sciences. I found the labs too tedious, and I had no intellectual investment or scientific curiosity for the laboratory exercises. I missed the purely theoretical aspects of mathematics.

Realizing I would be much happier if I switched my major from biochemistry to mathematics, I changed my major. Finally, I could really spend time learning more advanced mathematics beyond calculus and try to learn some physics. I fondly remember leaving math classes with great excitement regarding all of the mathematical ideas I was learning. As a math major, it felt absolutely wonderful. I was able to think about mathematics all the time. Professor Arthur E. Fischer and Professor Anthony J. Tromba, both highly influential professors at UC Santa Cruz, completely convinced me of the incredible beauty and versatility of differential geometry. Their lectures were very enthusiastic, riveting, and inspiring.

After graduating from UC Santa Cruz with a BA in mathematics, I stayed in Santa Cruz and worked for the UC Santa Cruz math department as a grader and a math tutor. In addition, I took a couple more math classes, and then I enrolled in the master's program at UC Santa Cruz. After writing my thesis and taking the algebra qualifying exam, I obtained a master's in applied mathematics. After completing my degree I took some time to consider the following question: Should I go further and get a PhD? During this time of contemplation, I returned to West Hills Community College because I accepted a position as an adjunct instructor there. It was also a great chance for me to thank the institution where I got my start. Had it not been for the existence of West Hills Community College, I would not be writing this article now.

## Am I Capable of Earning a PhD in Math?

**Earning a PhD.** As part of my decision on whether or not to earn a PhD in mathematics, I felt that it was important to go to a different institution so I could see how other places did mathematics. I wanted to study Einstein geometry under Professor Charles P. Boyer. Charles Boyer was one of the leaders in that area, having discovered a new technique of constructing special types of Einstein geometries. Einstein geometry is a type of geometry that obeys an equation discovered by Albert Einstein in his modern theory of gravitation-general relativity.

After numerous thorough discussions with friends, I decided to accept University of New Mexico's offer to enroll in their PhD program. Part of the attraction in attending UNM involved a generous stipend sponsored by the New Mexico Alliance for Graduate

Education and Professoriate (NMAGEP). This was a fantastic program that not only supported me with an additional stipend, but also provided numerous conferences and workshops for graduate students from underrepresented groups that focused on navigating the challenging road to becoming a professor. Looking back, this program was instrumental in helping me to think about what it meant to be a professor.

After postponing my fall enrollment at UNM, I arrived there in January of 2003. It was refreshing to be studying mathematics once again, and I was growing increasingly optimistic about my future career trajectory. But this optimism was cast into the shadows. In the early fall of 2003, it was determined my father had stage four colon cancer. By the time the malignant mass was found, it was too late for any procedure or radiation to prolong his life. He passed away in September of 2003. Days before he died, quitting the PhD program was weighing heavily on my mind. If I withdrew from the program, I could return home and help out my mother. My sister planned on moving back to Lemoore from Palm Springs with her family. I felt like I was abandoning my family if I remained committed to the PhD program. The afternoon before my father passed away, I was sitting next to him, telling him my final goodbyes. By this time he was extremely frail and life seemed to be visibly evaporating from him. But somehow my father was able to conjure a sentence: "Don't let this mess up your schooling." In that single sentence, the decision for me to complete the PhD was solidified. I simply had to finish now.

**PhD**. With my father's support and my sister's willingness to uproot her life to take care of my mother, I stayed in UNM's PhD program. Around this time, my advisor Professor Charles Boyer gave me a graduate fellowship from his research grant which allowed me not to teach and focus on courses and the beginnings of research. To have a professional and successful mathematician believe in me and encourage me the way he did, instilled a great wave of enthusiasm and excitement in me as I moved forward with my total immersion in mathematics. I earned my PhD in five-and-a-half years (with distinction) and a new life was ahead!

 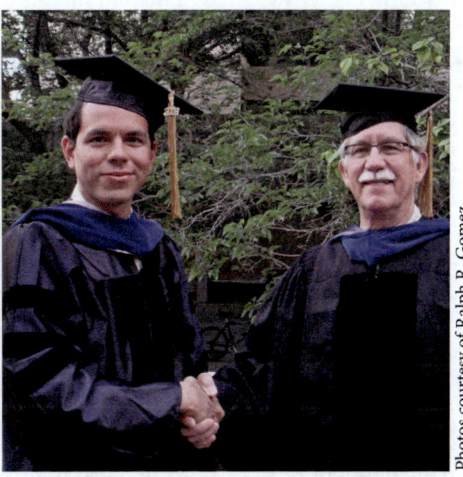

Photos courtesy of Ralph R. Gomez.

(L) My mother and me after my PhD ceremony, 2008. (R) My advisor Charles P. Boyer and me after he hooded me, 2008.

# The Professorial World

After I earned my PhD, I accepted a two-year visiting position at Swarthmore College in 2008. This was an absolutely incredible position with a reduced teaching load, and I was thoroughly excited to be there. But gradually I had the uncomfortable feeling that I did not belong at such an illustrious institution. Not seeing much faculty diversity at the college, particularly in the Natural Sciences and Engineering division, made me feel more like an outsider. It became clear to me just how different my pathway was in becoming a mathematician and that perhaps I was ultimately doomed to failure since my pathway was not more traditional. Impostor syndrome took a strong hold of me. To further complicate matters, my brother Johnny was on his way back to prison yet again to serve a one-year sentence. What other faculty member in their first year in a visiting position had to worry about a sibling heading back to prison? To my mind, this was just one of many other examples of why I did not belong.

Near the end of my two-year position, Swarthmore College was able to offer me an extension on my visiting position. This was very generous especially since this was around the time of the Great Recession. The college was actually pleased with my work and invited me to stay on for a few more years. This invitation was a clear signal that the college viewed me as thriving at Swarthmore. But there were three factors that led me to decline this offer at the end of my two-year visit. First, the desire to return to my family in California particularly because of growing worry about how my mother was handling my brother's return to prison. Second, the personal belief that I was not Swarthmore material and thus did not belong. Finally, the need to secure a tenure-track job instead of staying in a visiting position. Not feeling like I belonged at Swarthmore was the main factor that ultimately convinced me I should move on. Thus, after my two-year visiting position I left Swarthmore college and took a tenure-track position in California. Before I left I met with Stephen B. Maurer (the chair of the mathematics and statistics department at Swarthmore at the time) to tell him the main reason why I was leaving. He sent me an email after our meeting which  completely shifted my view about myself. With Stephen B. Maurer's permission, here is the email he sent to me:

> I'd like to address the worry that you were brave enough to broach with me today: are you really good enough for Swarthmore? It's really the same issue as when we admit students who have no history in their families of fancy colleges, or any colleges, or any history of expectations as demanding as ours or of positions of substantial responsibility in society. Swarthmore's belief is: people with the right underlying talent don't have to be brought up to the top gradually through several generations. They can leapfrog to comfort in an environment of high expectations in a few years. If we are right about this—and surely we are right in some cases—then Swarthmore, and not more laid back places, is really the place to make these people shine. This goes for the students from humble backgrounds that you inspire here, and it goes for you.

I never responded to Maurer's email because I did not know how to respond. It shook me to the core. I thought very hard about this email over the next several months as I adjusted to my new tenure-track position in southern California. I eventually came to the startling conclusion that Maurer was absolutely right! Moreover, I realized I was selling myself short. It is as if I finally let myself accept the idea that I did a successful job at Swarthmore. Within the first semester at my new position, I told Maurer that I was heading back on the job market. It turned out Swarthmore was able to offer me a chance to return but this time as a tenure-track assistant professor!

In 2012, I returned to Swarthmore as a tenure-track assistant professor and achieved tenure in 2017. My mother passed away in 2018 and so I am very thankful she lived long enough to see me achieve tenure. She was always one to express how immensely proud she was of me. Returning to Swarthmore College is without a doubt one of the best decisions I have ever made in my entire life. With the help of an extremely supportive department and surrounded by inspiring students, I finally feel that I belong at Swarthmore.

## Advice

Because my pathway to being a mathematician was not along a traditional path, I assumed that my role as a professor would therefore be less valuable and ineffective. This point of view was highly corrosive. It took a lot of conversations with colleagues, friends, and family to realize this view of myself was completely inaccurate. Something that I would like to impart upon the aspiring mathematician is this:

*There are countless paths to having a deep and meaningful relationship with mathematics.*

However, like any journey along an arduous path it is immensely helpful to have useful resources. I cannot stress this enough, dear reader. Build a network of people you can reach out to for help, advice, or direction. Seek out feedback, viewpoints, and opinions from others in your support system. You may be surprised how open people can be in giving you effective guidance. A favorite teacher, professor, friend, and family members are just a few examples of people you can add to your network of support. Even more support can be found for example through the Math Alliance (mathalliance.org), which has a vast database of professors who are ready and willing to mentor students interested in the mathematical and statistical sciences. With a solid support system at your disposal you will be able to be inspired and encouraged to carry on even in the darkest hour. And carry on you must for one day it may very well be *you* who takes up the role as a mentor!

# 14

# Dr. Victor H. Moll

## The Early Years

I am not sure when, but part of my family came to
the south of Chile from Germany as a nneteenth-
century immigration policy by the Chilean government.[1] What I do know is that in the 1950s it was
an established tradition in medical schools in
Chile that upon graduation, doctors would serve
in small towns before applying for jobs in Santiago
and other large cities. My father, Victor Hugo Moll
Strassburger, graduated in 1955, married Ema
Lucy Becker Correa, his girlfriend since his first
year of medical school, and took his first job in
Cabildo, a small mining town north of the capital.
In Chile you start medical school right after high
school, so they dated for a long time. He was the only doctor serving several small villages.

Dr. Victor H. Moll

Illustration created by Ana Valle.

My parents, 1954.

Photo courtesy of Victor H. Moll.

---

[1] In the nineteenth and twentieth centuries, Chile established immigration policies that encouraged European immigration.

Photo courtesy of Victor H. Moll.

My fourth-grade class, Cabildo 1964.

I am the oldest of three. My sister Ana Maria, my brother Ricardo Antonio and I were all born in Santiago, since the capital had hospitals with better facilities. My parents' families all lived in Santiago, so we often visited them. The few memories I have from that time always involve relatives coming to spend time with us, with my grandmother Clara Moll Strassburger directing the group. Those visits by relatives always involved lots of cooking and among my favorite sweets were *calzones rotos*,[2] which is a deep fried cookie full of powdered sugar. After my father's untimely death in 1963, my mother and the three of us stayed in Cabildo for one more year.

So I spent my early years in Cabildo, starting my formal education in *Escuela de Hombres, Número 5*. This was a typical elementary school in a small town, probably with students of different ages in the same room. My teachers were Angelina Guzmán and Maria Eugenia Palacios. The photo above shows my fourth-grade class, I am the fifth from right to left in the middle row. Through social media, I have been able to reconnect with some of my classmates and with my teacher Ms. Palacios who sent the picture.

After my father's early passing, we moved to Quilpué, a town near Valparaiso. The privilege of being the family of the town doctor had ended. My mother, then 33 and a widow with three young kids (I was seven and the oldest), had to learn how to survive. She remarried Sergio Labarca very soon after that. She used to tell me, that without a doubt, this was one of the best decisions of her life. My siblings and I gained a new father, in the complete meaning of the word. It was his opinion that education was the most important gift parents can give to their children. He found one of the best schools in the region (and one of the most expensive ones). Most of the small income our family was receiving, with both parents working, went to pay for education. Therefore, I started fifth grade at The MacKay School. My sister and my brother went to similar types of schools. This was a British school founded in 1857 to serve the community of immigrants coming from England as part of the business activities around the port. Before the opening of the Panama Canal, Valparaiso was a major port for ships going from Europe to the West Coast of the United States. It was here that my teacher, Maria Eugenia Pardo, noticed that I had some talent for mathematics. I still remember that she was very happy when I was able to show that any angle inscribed in a semi-circle was a right angle. This was seventh grade,

---

[2] This literally translates to "ripped underwear."

a period in which mathematical education in Chile was being guided by abstraction and axiomatic mathematics was taught even at this level. For me, there were some inherent life complications being from a working class family and being a student at a fancy private school. The economic standing, naturally associated from being the son of a doctor, had ended. Somehow mathematics became my refuge.

I finished my secondary education at the public *Liceo Coeducacional de Quilpué*. It was a very stimulating time: the country was going through very interesting social changes in the early years of the 1970s and being in high school at that time offered many experiences that built character. Many of my classmates left the country after the coup d'état and are scattered all over the world. We still get together via electronic gatherings, which sometimes have to be early in the morning in order to accommodate those with very different time zones. Some years ago I had an interesting experience when I was invited to give a talk about my academic path by the office of the mayor of Quilpué. The chance to give a presentation to high school students about my academic life made me uncomfortable. The school had decreased in quality and was beginning a slow recovery period. For many years, the Chilean public had been convinced by authorities that private schools are always better than public ones. This had the consequence of depriving public schools of funds needed to function, leading to a deterioration of what was a very good school at the time that I attended. The beginning of recovery began when the mayor's office decided to aim towards students interested in arts. During the time of my invited lecture, I met with many students, and tried to convince them that it is possible to be interested in science and not fit the stereotypes (they assume that if you liked mathematics, you had to be a nerd). I was lucky that two of my best high school friends, Kenna Meneses and Juan Francisco Carrasco, came to my presentation and vouched for my stories. At the end, they seemed to like what I was telling them. Although, I had returned to my high school with mixed feelings, having the opportunity to talk to the students made it all worth it in the end.

## Undergraduate Studies

After graduation and with the knowledge that the best option for a high school student with interest in mathematics was to join an engineering school, I did so. In March of 1973, the beginning of the fall semester, I began my studies at *Universidad Técnica Federico Santa Maria*, one of the most prestigious engineering schools in Chile. The core part of the curriculum was common to every student, including three mathematics courses. It was there that I realized that my background was not optimal. Many other incoming first-year students had seen calculus in high school. This was all new to me. And then came the coup, September 11th, 1973. Learning integral calculus with a curfew was challenging. All of my undergraduate education was during the new regime.

The geographical isolation of Chile, coupled with a nineteenth century immigration policy that allowed only Europeans mostly from England, Germany, and Yugoslavia to immigrate, created a more homogeneous society than some of our neighboring countries. I am not sure of all the details, but I believe that this is how my ancestors came to the south of Chile. This centralized immigration policy and the lack of travelers from other countries produced a relatively racially homogeneous population. I have no early memories of African, African-American, Asian, or other immigrants being part of our town. To me,

Photos courtesy of Victor H. Moll.

*Liceo Coeducacional de Quilpué* in 1971 (left) and 2014 (right).

the racial distinctions were weaker than the economic ones. More than that, the concept of class was very strong. There are even Chilean terms to describe the distinction between having class versus having money. Growing up, I never felt discriminated because of racial issues.

Although I took mathematics courses at *Santa Maria*, I began my career in college on a track to become an electronic engineer. In the university educational system in Chile, you choose a career at the moment you enter college. There is no concept of having a major. When I arrived, studying mathematics had been closed as an option to all incoming students, even though this had been an option in previous years. Fortunately, in my third semester as an undergraduate student at *Santa Maria*, I managed to transfer to become a mathematics student. There were only two other students in the mathematics program.

There are many differences between the Chilean and American university systems. From the point of view of this story, the most important one is the fact that in Chile students choose a career at the end of high school. If you want to be a lawyer, you go directly to law school. No time to warm up. If you do not like it or do not do well, you have to retake the entrance exams and apply again. If you are a student with some talent in mathematics, the most natural choice is to go to an engineering school. So I did. For some bureaucratic reason, at the end of my second year, I was allowed to transfer internally. This saved me from applying to university again—my parents would have been supportive, but not happy if I had to start again. The mathematics degree was a five-year program and during the last three years it had only three students. We had mostly mathematics courses, some courses in English (my father's plan to put me in a British school paid off) and once in a while we registered in some physics courses. Classes were obviously small, sometimes in the instructor's office. This is where I learned some analysis and to like espresso. There were lots of independent studies courses. Essentially, it meant that the instructor would choose a book and the students would lecture each other. During some semesters we would take classes at the nearby *Universidad Católica*. This gave us a chance to learn material not offered at Santa Maria.

This was a period of transition in the life of the country. Among the instructors, there was a single PhD in mathematics, which was unique mostly because the university was not in Santiago: the center of everything. He was the renowned Roberto Frucht, an expert in graph theory. At a moment where I was thinking of abandoning mathematics and studying something else, a second PhD came to the department. Luis Salinas C. came back from Germany, rescued me back to the subject that I loved and became my advisor. Many times in my life I have been lucky and having him return to Chile then is among these lucky times. My last three semesters I took all of my courses with him. I owe him more than words can say.

Things have changed in Chile since my days as an undergraduate. Many Chilean mathematicians came back to the country. They have created a wonderful educational structure and Chilean students travel to the best institutions in the world to study and many renowned mathematics departments have faculty from Chile.

## Graduate School

Upon completion of my undergraduate studies I was hired as a faculty member of the *Departamento de Matemáticas of Universidad Santa Maria*. At that time (1978) it was not required to have a PhD to teach in a university. Yet my undergraduate thesis advisor, Prof. Luis Salinas C., was always talking to me about going abroad for a graduate degree. A real opportunity to go abroad developed with the visit to *Universidad Santa Maria* of Prof. Eugene (Gene) Trubowitz from the Courant Institute of New York University in July 1980. Since I spoke English (having been a student at The MacKay School) it became my role to be in charge of his visit. I still remember walking on the beach, with conversations that usually started as "Victor, suppose $A$ is a normal matrix of size $n$." To make a long story short, I joined the PhD program at NYU in September 1980 with financial support from my school. Since I was late in the application process and had not been aware of the required forms, I went to the American Embassy in Santiago with a telegram from Gene that essentially said "Come to New York, we will fix the paperwork here." As you can imagine, the first time I showed up at the embassy carrying only this telegram I was denied a visa. The next time, it occurred to me that if I spoke English to the guard my chances might improve. They did. I got a tourist visa and left for New York. The paperwork was fixed after my arrival.

It is hard to describe my early days in New York City. I came from a relatively small town without much foreign influence, where everybody looked like part of the same family. This was the time before the 1985 economic boom in Chile, when the country essentially became a 51st state. I made many mistakes in those early days. Perhaps the worse one dealt with housing. NYU owned a group of buildings nearby and studios were assigned to incoming graduate students. The first time I went to the housing office, they told me about this option and I realized that rent was about 40% of my total income. Immediately I refused the offer, much to the consternation of the person in charge. I did not understand why she kept explaining to me that these studios were my optimal choice. Needless to say, my off-campus living accommodations during the first year of graduate school ended up being inferior. Someone should have grabbed my hand and told me to sign the dotted line. It takes time to learn the American system.

The schedule at Courant was such that classes would meet once a week for two hours. For me this meant that at the beginning of being in New York, before making friends, I had no human contact from Thursday night until Tuesday night. It came as a great surprise when one day after class, in the elevator going down, this person said to me, "We seem to be in two classes together, would you like to have a cup of coffee?" He was Fred Schiafando, who turned out to be a friend for life. Lectures at Courant moved fast, and soon a group of students decided to get together to study. Social life improved from that point on.

Coming from an educational system with lots of classes I was surprised when at registration I noticed that taking four classes meant eight hours of contact. Naively I asked, "What I am supposed to do with the rest of my time?" The response was absolutely correct: "Try to catch up." I was lucky to have Prof. Henry McKean as my instructor for complex analysis. His style of lectures and his point of view of mathematics as a whole made a profound impression on me. Being his PhD student has been one of the biggest honors of my academic life. At the beginning of my second year, I met a first-year graduate student: Lisa Fauci, who later became my wife. My life with her has been great since then.

## Professional Career

After graduation, I took a postdoctoral position at Temple University in Philadelphia. Recently, cleaning my office, I found a copy of my job application: it was one-and-a-half pages long. It is remarkable how things have changed. During the next two years I spent a lot of time on the trains between Philadelphia and New York. I also attended a class on elliptic functions that Henry McKean gave at Courant. I worked out all the possible details and years later Henry and I coauthored a book.[3] In 1986, Lisa and I applied for jobs together and both took jobs at Tulane University in New Orleans, Louisiana. Our unspoken plan was to be in New Orleans for a short time and then move back to the Northeast. We never left.

First off, New Orleans is a great place to live in. New Orleans is not your typical American city. It is humid. Most of the time things do not work properly. Sometimes, the driver of the street car stops the ride so they can get themselves a cup of coffee and will turn around and will tell you to take your feet off the seats. Music is one of the city's biggest priorities and soon became always present in our home. My kids, Alexander and Stefan, had a chance to become students at NOCCA (the New Orleans Center for Creative Arts) as jazz musicians. You should come visit, but come first when there are no big parties happening, which does not leave too many open days! One of my favorite events in the city is Jazz Fest, it takes place during two weekends in April-May. It was a fantastic feeling watching my son Stefan play piano on the Blues Stage. Such is life in New Orleans.

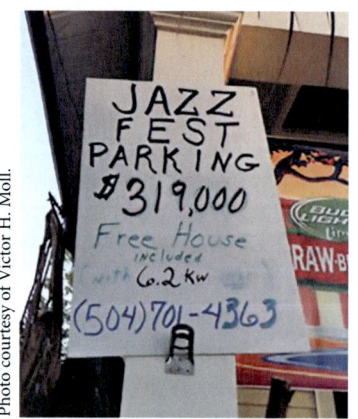

Photo courtesy of Victor H. Moll.

New Orleans humor.

---

[3] Henry McKean and Victor Moll, *Elliptic Curves: Function Theory, Geometry, Arithmetic*, Cambridge University Press, 1999.

Photo by Paula Burch-Celentano.
Used by permission of Tulane University.

Weiss award photo, New Orleans, 2011.

It has been great to be at Tulane University. Among the many positive aspects of working at Tulane, I must say that the existence of a great child-care program, with a dedicated group of teachers that educated our kids while we worked in peace, is one of the best benefits that I have had. In addition, the undergraduate students that I have had a chance to work with have made my job a very enjoyable one. Among the many students, I would like to single out Roopa Nalam, who worked on a research project with me involving integrals of special functions, and Kirk Soodhalter, who wrote a thesis on some interesting aspects of modular forms. Roopa went on to complete an MD-PhD program and now is an Assistant Professor at the Baylor College of Medicine. Kirk completed his PhD in mathematics at Temple University (in numerical linear algebra) and is now an Assistant Professor at Trinity College in Dublin. I am very proud of the two of them. These two students are part of a large group of students who have enriched my academic life. Some years ago I was awarded the Weiss Presidential Award for Graduate Teaching. Since this award is chosen among faculty nominated by students, this had a special meaning to me.

## Research

**Working with students and young faculty**. Over the years I have also had opportunities to work with undergraduates outside of Tulane. One day, I was asked by my friend and colleague Prof. Ricardo Cortez if I wanted to go to a conference of the Society for the Advancement of Chicanos/Hispanics and Native Americans in Science (SACNAS) in Portland, Oregon. I had to admit that I had never been aware of this organization. For me, one of the highlights of the conference was to meet Prof. Ivelisse Rubio and Prof. Herbert Medina. After I gave my talk, they told me about the Summer Institute in Mathematics for Undergraduates (SIMU), an undergraduate Research Experience for Undergraduates (REU) program at the University of Puerto Rico at Humacao, and asked me if I was interested in being the senior leader. This meant being in charge of 12 students doing research for about seven weeks. I still remember their voices telling me "This is a

MSRI-UP research group, Berkeley 2014.

lot of work." I accepted. My family and I spent a great time in Puerto Rico. They were at the beach and I was working. Even though the workload was enormous, I still believe that the SIMU model is one of the best for undergraduate programs. The students were exposed to research-level mathematics, which opened new avenues for them intellectually, but also were told how to approach the graduate school application process, how to present a paper at a conference, what is expected of them as members of the community, and many other aspects of being a mathematician. I loved it. Luis Medina, now professor at the University of Puerto Rico at Rio Piedras, was my student at SIMU. After that, I was lucky that he chose to come to Tulane University for graduate school. We still maintain a scientific collaboration. I am very proud of him. The SIMU model was later adapted by the Mathematical Sciences Research Institute (MSRI) at Berkeley, California in their Undergraduate Program (MSRI-UP). Participating there was a wonderful experience again, as I was surrounded by intelligent students and assisted by graduate students and postdocs. Many of the students participating in this program are now faculty members at a variety of schools.

In recent years, I have been involved in wonderful programs whose mission is to engage faculty—who work in schools with a high teaching load—in research. The programs that I have participated include PCMI (at Park City, Utah) as part of a special program in Random Matrices and the second one at ICERM as part of REUF (Research Experiences for Undergraduate Faculty). These are exceptional programs and I would like to encourage the mathematical community to participate in these programs. I am certainly planning to continue doing it.

**Math interests**. My mathematical work these days started when a former graduate student, George Boros, told me that he could evaluate an integral. I have told this story in detail in "The evaluations of integrals: a personal story," *Notices AMS* (2002) 311–317 and

Research Experience for Undergraduate Faculty (REUF) research group, Rhode Island 2018.

"Seized opportunities," *Notices AMS* (2010) 476–484. George's comment transformed my research. In fact, the evaluation of definite integrals will take you into many interesting areas of mathematics, and you should try it.

The basic question is this: given a function $f$, defined on an interval $[a, b]$ (with $-\infty \leq a < b \leq +\infty$) one wants to evaluate $\int_a^b f(x)\,dx$ in terms of special values of a class of functions. This is familiar from elementary courses. On the other hand, students are often told that the function $f(x) = e^{-x^2}$ does not have an elementary primitive, but there are methods to see that

$$\int_0^\infty e^{-x^2}\,dx = \frac{\sqrt{\pi}}{2}.$$

As a freshman, I was told that this exponential function does not have a primitive. I vividly remember my reaction: "*Maybe you do not how to integrate it, but I will figure it out.*" Needless to say all of my efforts went nowhere.

In one of my luckiest academic moments, my former student George Boros told me that he could evaluate

$$\int_0^\infty \frac{dx}{(x^4 + 2ax^2 + 1)^{m+1}}$$

in terms of some elementary radical and a complicated polynomial $P_m(a)$ in the variable $a$. He had a mechanism to evaluate the integral of a rational function by doubling the degree of the integrand and, by some symmetry reduction in certain special cases, he was able to cut the degree in half. That gave him a complicated expression for $P_m(a)$.

At the beginning, not being particularly impressed with evaluating integrals (prejudice comes in many forms), the only part that I thought might be interesting was that his procedure vaguely reminded me of Landen transformations for elliptic integrals. I started trying to evaluate this example by other methods and failed. My not knowing about hypergeometric functions ended up being a good thing. One can prove George's result in an elementary manner using these functions. The work on this integral took us through an almost magical trip visiting many parts of mathematics which appeared to be disconnected. Many roads of this mathematical trip are mysterious. For instance, the integrals above are the (Taylor) coefficients of $\sqrt{a + \sqrt{1 + c}}$, expanded as a power series in $c$. Very strange. George's proof uses a theorem of Ramanujan and to this day, I do not know how he thought of this. Since George passed away in 2008, the mystery will remain. For a long time I have looked for a different proof, using this double root. It should teach us something new about this problem. *So far I have not had luck in this regard.* It is needless to say that my original impression of George's problems was wrong. I am glad that I did not dismiss him right away and decided to pay attention to his methods. My current students continue to surprise me like this. I do my best to listen to them. It is one of the privileges of the business we are in.

# 15

# Dr. Ryan R. Moruzzi, Jr.

## My Story Begins with my Mother Irene

My mathematical path was influenced by my parents, especially my mom, Irene. My mom's dad and my grandpa, Raul Yzaguirre, was in his early twenties when he was diagnosed with myasthenia gravis, a neuromuscular disorder that affected his ability to walk and breathe. Doctors told him he would not live long, and that he would never walk again, forcing him on disability. He used a wheelchair for most of my mom's early life, so from a young age, she helped care for him.

Illustration created by Ana Valle.

Dr. Ryan R. Moruzzi, Jr.

Throughout my mom's childhood, my grandpa frequented various hospitals, and when not in the hospital, my mom would help care for him with his endless daily medication, setting alarms for him at night so he could wake up and take them. The associated financial and emotional struggles of caring for my grandpa impacted my mom in a way that made her conscious of the stress her parents were going through, and therefore, she always pushed herself to do well in school.

My mom was one of six kids, and throughout high school, she balanced caring for my grandpa, working, and studying. She gave her paychecks to her parents, and overall she helped out in whatever way she could. My grandpa always told my mom, aunts, and uncles to do their best in getting their education, and when my mom heard him say this, she knew she had to do her best.

After high school, my mom first attended San Bernardino Valley College, a community college in San Bernardino, CA. Part way through, not finishing at Valley College, she switched to a program at Loma Linda University in Loma Linda, CA, to become a licensed vocational nurse. My grandpa told her she would make a great nurse because of how she would always take care of others, especially helping him with his various tasks throughout the day. She had a discouraging experience during an internship where she was told that she would *never* make it as a nurse, that being a nurse was not for her, because she was not doing some tasks correctly. This was damaging enough to make her drop out of the program. She did not feel the confidence to continue. She did not complete this program,

Irene standing with her mom, Frances
Yzaguirre, when I was first born.

and she did not return to school. It turns out,
her academic journey was very similar to
and influenced my own.

# K-12 Academic Experiences

My K–12 journey was injected with mathematical confidence early on, brought about
by circumstance, and questioned at every
step of the way, which was not unlike my
mom's own experience while in school.
My schooling was done through the Rialto
Unified school district in Rialto, CA. Rialto
Unified was and still is, a Title I school
district.[1] While I was in elementary school,
my mom realized that I enjoyed things that
challenged me intellectually; she sought
to discuss how the school could push me
academically. Because of her persistence,
various opportunities arose, including being placed in a program for gifted and talented
students, which helped me grow my confidence in school at a young age.

After elementary school , I attended Jehue Middle School. Jehue focused on STEM and
college tracks through GEAR-UP.[2] I recall visiting local colleges, such as California State

Figuring a puzzle in elementary school.

University, San Bernardino, and participating in other programs to broaden college
readiness. Though my parents didn't graduate from college, I was consistently getting
the message of attending college from both
my school and home. At the time, the conversation was centered on getting a bachelor's degree, not really graduate school. In
fact, throughout my undergraduate career,
I was still unaware of the opportunities for
graduate education.

At Jehue, I started to gain more confidence in mathematics. Towards the beginning of my sixth-grade math class, we were
covering addition, subtraction, multiplication, and division of fractions. We had

---

[1] The U.S. department of education defines *Title I* as providing financial assistance to local
educational agencies (LEAs) and schools with high numbers or high percentages of children from
low-income families to help ensure that all children meet challenging state academic standards.

[2] Gaining Early Awareness and Readiness for Undergraduate Programs was designed to increase
the number of low-income students who are prepared to enter and succeed in post-secondary
education.

My friends and I (third from the left) posing with creative hats.

already discussed these topics in the previous grade, so I did not understand why we were discussing it again. Maybe my boredom in the class stemmed from the curriculum, or maybe my teacher did not recognize or realize students in that area could/can be challenged and pushed more than how they were. My mom was conscious and worried about keeping me challenged.

During this time, my grandma was working towards her associate's degree at San Bernardino Valley College. I note that my grandma did not graduate high school. She stopped going to school while in middle school. She married my grandpa at the age of 14 and spent her life caring for others, including her six kids. In the 1990s, she decided she wanted to get her General Educational Development (GED),[3] and eventually her associate's degree; she wanted to prove she could complete a college degree. One of the last classes my grandma needed for her associate's degree was her math class, college algebra. She had felt incapable in math, and struggled. My mom said I could help since I was good at math, and I looked at her workbook. I remember looking at this old workbook that was in typeset font with problems such as:

$$\text{Distribute and simplify } y^2z^4(yz + 2y^2 - 3z) + 4y^2z^5.$$

Reading through some of the book, I figured the problem was just a matter of following some rules or guidelines.

After helping my grandma, my mom approached my middle school counselor, Mr. Ed, with the work I was doing. The discussion was centered on how to further challenge me. They decided to pursue moving me up to the next grade in only math. Along with my mom, Mr. Ed became one of my biggest advocates. He wholeheartedly supported the move for me, and with my mom, they further advocated for me in discussions with other teachers and administrators.

---

[3] The General Educational Development tests are a group of subject examinations which when completed are equivalent to the U.S. high school diploma.

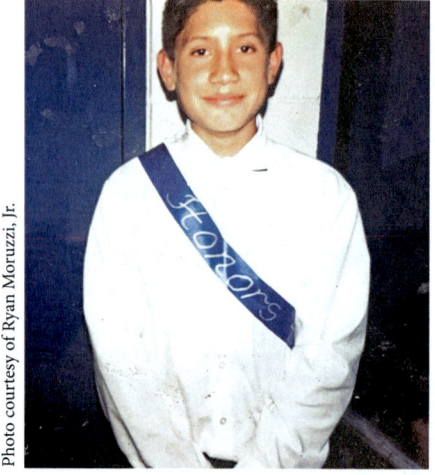

Jehue eighth-grade honor roll.

To prove I was capable of doing mathematics at the next grade level, I was taken aside into a room by two math teachers: one male and one female. For some reason, I remember the male teacher being more stern and interrogative, while the female teacher, was more inquisitive. I was given a test in a classroom, then asked to explain my reasoning on various questions. After the diagnostic test, they did figure that it was me doing the math, and I had the "correct" thinking with the problems and the only remaining discussion was on the logistics to accommodate me in the next grade level. Luckily, at the middle school, it was easy enough to bump me up to pre-algebra, which was the seventh-grade mathematics class.

In the pre-algebra class, I was pointed out for being the sixth-grader. The teacher would often call on me as though I should have answers to all of the math questions. The teacher acted like I should be able to do anything thrown at me. I would answer the best I could, and this further helped me develop confidence in my mathematical abilities. Even when I got some part of the answer wrong or if I had to think more about the problem, I still answered. I was viewed by my peers as the one to ask math questions to. This also boosted my confidence since I saw myself as being further along in mathematics than my peers. I now realize that through the move from sixth-grade math to seventh-grade pre-algebra, I gained the confidence that would later propel me to choosing to be a math major.

In my eighth-grade year, I had my last class of the day, geometry, at the high school that was about half a mile down the street from the middle school. Every day, after my fifth class, I would spend fifteen minutes in Mr. Ed's office, talking with him. He would check in with me about my classes and how things were going. I would then leave the middle school and make the walk. I was emboldened with the responsibility of making that trip every day, walking onto a high school campus before my peers. In a recent conversation with Mr. Ed, with whom I still keep contact, I found out that he would drive in his car and watch me on that walk every day; he made sure I would make it to the campus. I was shocked to find this out! I always thought I was on my own. When I found this bit of information out, withheld for so long, it made me well up with emotions. I still have trouble putting my feelings into words: thankfulness, pride, endearment, self valuation, belonging, cheerfulness. I found this out at a time when I was pondering what things along my path encouraged me to pursue a PhD math program, and that single circumstance of me walking and feeling empowered aided in that.

Throughout my early school education, Mr. Ed was an advocate that helped propel me towards becoming a mathematician. Along with my mom, his voicing support for me to move ahead in math, even taking the time to watch me walk to the high school, was work to ensure I was being challenged in mathematics. I don't know if I would have chosen math without his and my mom's support. After I left Jehue to attend high school, the middle school created a geometry class for students who were ready for such a challenge. I feel proud that, because of what I was able to do in mathematics, other students were being challenged and encouraged to tackle more advanced mathematics at a younger age.

## Journey towards an Undergraduate Math Major

My undergraduate experiences parallel my mom's, with the difference being the choice I made after a discouraging encounter with a professor. Growing up, I saw my parents struggle financially and, like my mom, this struggle made me conscious of the importance of my successes in school. Although my parents were proud of my every step and achievement, they could not give me guidance on navigating college. GEAR-UP and other programs helped in this regard by introducing me to colleges and by keeping the message of attending college visible.

Leaving high school, I applied to various schools and got accepted to two. I chose to attend Cal Poly Pomona because of my desire to be an architect, which I credit to growing up with imaginative play and building different structures out of Legos. Also, since I was comfortable with mathematics, a math-based career felt natural.

From the onset of college, I had to adjust and adapt. When I received my acceptance letter to Cal Poly, I learned I was accepted into the program of electrical engineering. This was my second choice since we had to choose more than one potential major in our application. I did not know that architecture at Cal Poly was an impacted major, meaning that I had to not only apply to Cal Poly, but also to their specific program. Therefore, I decided to stick with engineering because it was still math-based, and it gave me an opportunity to make good money, even though I must admit I did not exactly know what electrical engineers did.

In my first year of undergraduate education, I enjoyed and did well in my classes, riding a natural wave of classes and homework. In the fall semester of my second year, I took a digital logic class with a lab where we built various electrical circuits. In this class is where, similar to my mom, I had a horrible discouraging academic experience that greatly impacted my academic trajectory. A difference here was that I had the confidence to fall back into another subject, and that confidence stemmed from my mom and my K–12 experiences in mathematics. In the engineering lab, we were supposed to collect and put various resistors and capacitors together to successfully build circuits each week. I was left to my own devices to figure things out. I would spend hours and hours before and after class trying to figure out how to build the circuit we were supposed to build. I would ask the professor, but he would scoff at me and tell me I should be able to figure it out on my own. The grade centered on completion of the labs, and going into the fourth week, it was clear that I was not able to do them on my own. With no guidance, I would fail the course.

When I had that negative experience, I believed I did not have the skills or capability to become an engineer, and I needed to change my major. Not knowing what other major to switch to, coupled with the fact that I wanted to graduate in four years, I switched over to being a math major, which was not viewed as an "easy" major. I had confidence in math, and that confidence came from my mom advocating for me in those early years and other early experiences. This made me think about what would have happened for her if she had confidence in some other area, or had someone to advocate on her behalf. Switching to a math major meant I got some push back from some of my family because I decided I

My grandma and me at Cal Poly graduation.

would teach high school math; at the time it was all I knew that one could do with a math degree.

As a math major, I went through my classes with a closed mindset. I would push through knowing the end goal was to go off and start teaching high school. My desire to become a teacher was reinforced through my work as a math tutor in GEAR-UP at La Puente high school in Hacienda Heights, CA, and as a math tutor for Sylvan learning center in Rialto, CA working with underserved K–5 students. I enjoyed being able to help students with their perceived struggles and also being a mentor for them, planting the seeds of attending college.

In my own math education, there were definitely classes in the major that I struggled in, including both real analysis and abstract algebra. Both real analysis I and II were difficult for me. I never really felt as though I understood the concepts, and the professor never seemed concerned. But, in abstract algebra, there was an instance that steered me towards more of a comfort with algebra as a topic of interest. In the fall of my third year at Cal Poly, my first abstract algebra class was poorly run and I did not come out of that class with much knowledge, leaving me unprepared for the next course in the sequence. I also rarely attended professors' office hours, and I was worried about taking abstract algebra II (rings and fields). I sought out advice from the professor teaching that class, Dr. Robin Wilson. He reassured me that everything would be good, and he said that I did not need a great background of groups to be successful with rings and fields. That little moment of affirmation kept me on track. Throughout his course, he reignited my interest and confidence in algebra, which later led me to studying representation theory in graduate school, though I didn't know it then.

## Journey towards Graduate Mathematics

Leaving Cal Poly as a math major, I had one goal in mind—getting my teaching certificate. I moved to Michigan with my wife (then fiancée) and was on track to attend the University of Michigan's School of Education. I was planning to complete a bachelor's degree in education along with gaining a certificate, which was naive on my part because I only needed my teaching certificate. I was unaware that the government's grants, such as the Pell grant, only help you fund one degree. Not having money for out-of-state tuition to become a high school teacher and still not knowing what to do as a math major, I quickly pivoted to pursue a master's in the mathematics program at Eastern Michigan University. I was still leaning towards teaching, yet the graduate students at Eastern Michigan only graded or worked in the tutoring lab on campus. Through various discussions with the interim chair, Dr. Carla Tayeh, a fellow graduate student and I were able to pilot a program enabling graduate students to teach lower-division math classes. This was the first instance where I advocated for myself, and I got a taste of teaching at the collegiate level.

I already felt comfortable teaching, and doing so in college was affirming my growing thoughts of staying in the college classroom. This was another moment that could have steered me away from a path towards becoming a professor. If Dr. Tayeh had rejected my opinions of allowing graduate students to teach rather than grade or tutor, then I am not sure I would have been steadfast on teaching at the college level. The openness and willingness of Dr. Tayeh to take a chance on me was another moment of affirmation along my mathematical journey.

Based on these experiences, I began to research what it would take for me to become a professor at a four-year institution; I did not know much about doctorate degrees. I applied to four schools, three mathematics programs, and one mathematics education program, all in Southern California. I also applied to teach as a part-time instructor at various community colleges. Our plan was to move back to Southern California. Either I would attend graduate school, or teach at a couple schools. I received three rejection notices. With our move back to California and still no word from the fourth school, I did not think too much of it; my mind had already switched to trying to find work. Then, in May of 2013, I got a call from the University of California, Riverside (UCR) asking if I was still interested in their graduate program. That said, I started graduate school fall of 2013, not exactly knowing what I was heading into.

When I got to graduate school, the struggle got real. I felt woefully underprepared for all of my classes. I remember thinking I was not able to do it because I did not put in the time before as an undergraduate or as a master's student. I realized I was in classes with people who did much more studying than I did. This realization turned fruitful because I quickly formed study groups with them and became a better student. The more I struggled and thought about the material, the deeper my understanding became; I realized the struggle was good and advantageous. Previously, I rode that wave from the attention I received in sixth grade, believing I was always "good" at math. Failing my first graduate exams in algebra and topology, others in my cohort seemed to be in the same boat as I was, and this was a saving fact. I realized we were all going through this process of learning, and supported each other throughout it; another moment of affirmation. Without this support and friendship, I would not have lasted long in the program. The community that was built early on definitely supported my successes, and still supports my successes today.

Throughout graduate school, I relied on my cohort for support with classes and began to find my way as an academic. In my third year of graduate school, post-qualifying exams (comprehensives), I distinctly remember a conversation I had with my advisor Dr. Vyjayanthi Chari that steered me down the path I am on now. I am not sure if she knows the impact that conversation had on me. It was at a 2016 conference at the University of North Texas. At the conference dinner, she and I began talking about jobs that students want after leaving graduate school. She posed various questions to me, sorting through her own thoughts, and made me think about what job I would want. For example, she asked (paraphrasing) "*Why are students getting or not getting certain jobs when they graduate?*". We had a PhD candidate graduating from North Texas that spring sitting with us who had accepted a tenure-track job offer at a four-year institution, and Dr. Chari asked the student what steps they took to get their job. The message from the student was involvement: she was involved in organizing conferences, seminars, attending various conferences, etc. The follow-up question to me from Dr. Chari was "*Why or why aren't our students doing such things?*", with a follow-up comment, "*The job market is a difficult thing to navigate.*" I thought much about that conversation and about exactly what job I would want leaving graduate school. Specifically, I thought: what was my end goal? I always enjoyed teaching, and working with undergraduates, so I knew I wanted to be at a four-year institution with a focus on teaching. I also wanted to become a professor at a four-year institution to continue to mentor students and hopefully be a source of inspiration for others that may question themselves as they complete their studies. From that conversation with Dr. Chari, I sought out ways to set myself apart from other job applicants, looking for opportunities to develop professionally towards that goal, not just through mathematics. I had always been interested in teaching and outreach, and that conversation with Dr. Chari empowered me to seek out such opportunities, unlocking a door that had been previously invisible.

The spring of my third year of graduate school was when I first started to get *involved*, and began to more actively advocate for myself. I got involved in various activities inside and outside the department at UCR, such as joining math club and attending various conferences and workshops that were offered; always intently looking for different opportunities. I never sought out such things as an undergraduate or in my master's program, but as a PhD candidate, I finally realized I was more capable and turned proactive instead of passive. I passionately *pursued* things I was interested in. I formed a reading course for undergraduate students in Lie theory, which was not done before in the department. I led work with the math department to organize a math circle type program in the Rialto Unified school district. I also led work with others to organize a seminar on equity and inclusion for math graduate students and faculty. These activities helped me take steps towards my end goal of becoming a professor at a four year institution. They also challenged my thinking about how to unlock students' potential. How can we *empower* and *support* students to pursue things that will help them reach their end goal? How can we, directly or indirectly, positively play a role in a student's path?

## My Family

Part of my *testimonio* and journey through academia is about my family. Having kids in graduate school was a decision that my spouse and I came to and wanted. I was met with resistance from some, receiving questions like, "Why now? Are you going to be able to keep up?" Some thought that if we had children I would not finish my doctoral degree.

 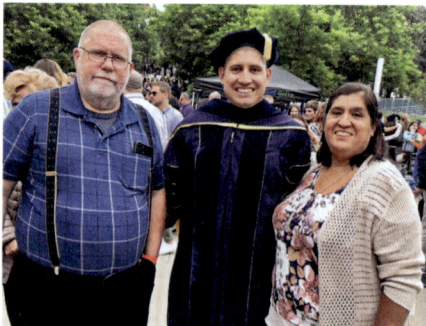

(L) My mom, Irene and grandma, Frances. (R) My dad, Ryan Sr, me, and mom, Irene.

L to R: Ryan III, Ryan, Jr, Bree, Reid.

Those questions and comments fueled me even more to prove to others that it can be done, though there were, and still are, private moments I question myself. Those moments of creeping self-doubt would then be backed by a thought of, "Why should the profession be void of life's experiences?" Yes, it takes extra work on my part, and it takes support from my spouse, my mom, my grandma, and others in my family. Part of the Hispanic culture, at least for me, has meant receiving unquestioned support from those around me, and that support continued throughout graduate school, still continuing today.

Being passionate about what I do, teaching and studying mathematics, made it easier to balance life and work. I gained the appreciation for the time I have to work and the time I have for my family. The tug-of-war between work and life is never-ending, always causing moments of stress through many moments of joy. Overall, the experience has been enlightening and rewarding.

## Concluding Remarks

I didn't take a traditional path to be a professor. In fact, my path towards the professoriate was laced with instances of advocacy and affirmation. My parents did their absolute best in steering my siblings and me with the resources they had. The choices my mom made in her early post-secondary career, leaving college and not finishing, turned into support and motivation for me on my academic journey. My mom's lived experiences turned into her supporting me by ensuring I had access and confidence in academics, speaking up to teachers to challenge me in school, and advocating for me and instilling a sense of self-belief. I was also motivated by her experiences which gave me a deep desire to make my parents proud, get to the finish line of college, and prove to others that I can do it. That support and motivation eventually pushed me to successfully pursue a doctorate in mathematics. Outside of that, programming in my K–12 schools was also crucial in supporting my thoughts of continuing education beyond high school. That is to say, one way towards supporting more Latinx/Hispanic people in mathematics is to be an advocate and provide support in multiple ways; small things can make a huge impact on a trajectory of a student, for better or for worse. This is what I carry with me throughout all my work. You never know what instance may positively influence someone on their path, and it is my hope that the instances I have shared in my *testimonio* will positively influence you on your path.

# 16

# Dr. Cynthia Oropesa Anhalt

## A Story of Latinx

Something I heard as a child from relatives was the story that God first "baked a batch of people," and the oven was not hot enough, so the people came out "underbaked" with light skin, so God compensated in the second batch by turning the oven temperature too high and burned the second batch creating the dark skin. For the third batch, God got the temperature exactly right, and created the brown skin Latinx people. This story warms my heart in that my extended family took pride in brown skin, but at the same time, the story is about competition and divisiveness, which has connotations of discrimination inside and outside the Latinx community. This story did not hold meaning for me until I became more aware of our ethnicity when I attended mostly-white public schools.

Dr. Cynthia Oropesa Anhalt

Illustration created by Ana Valle.

## Early Life and Immigration

My father, Sergio Duarte Oropesa, was born in 1935 in Ciudad Juárez, Chihuahua, México to parents who worked in customs for the Mexican government. By his teenage years, his family had lived in Mexico City and various Mexican-U.S. border cities, including Nogales, Sonora, Mexico, where he met my mother, Francisca Orozco. My mother was born on a ranch in Saric, Sonora, Mexico in 1933 and only completed up to third grade in school.

My parents married in 1956 and had four children while living in Nogales. In 1959, my father gained sponsorship from Selby Motors, a Mercury and Lincoln car dealership, to work as a car technician in the U.S. For nine years, he crossed the U.S.-Mexico border daily in a 1953 Chevy pickup truck, which he fondly remembers as his first vehicle. He remembers the assassination of John F. Kennedy and the national border closed for the day, forcing him to spend the night at a hotel in the U.S. while the family was across the border, which was the first time that my father was apart from the family. It was not until 1968 that my mother, my siblings and I immigrated. My sister, Luz Elena was 11 and has

The family in 1975.

a memory of seeing a large portrait of President Lyndon B. Johnson as we entered the U.S., while my brother, Sergio Agustin was 10 and remembers being in a 1964 Mercury Montclair sedan as we crossed the border. My next older brother, Moses, was six, I was three, and we have little memory of entering the U.S.

## Bilingualism and Biliteracy

Speaking two languages, Spanish and English, was a natural and organic part of my life. It seems that an unspoken rule was to speak Spanish at home with family and at church, while English was reserved for school and other events outside of the home. This practice became a form of diglossia, which became my parents' mantra for the two languages we spoke; each language played a role in different social contexts for performing different functions. This ease in separating the two languages was something I never questioned because of my parents' beliefs that they were responsible for teaching their children the heritage language. My father took English-language classes in the evening twice a week, and remembers proudly that he was the only student who was left at the end of the term. He ended up becoming close friends with his teacher, Señor Garcia, and continued taking private classes with him at no charge. At home in the evenings, my father would begin teaching English grammar to anybody that would sit by him long enough.

Conducted completely in Spanish, we attended church, and this became the space in which I learned to read academic Spanish language through the King James Bible. Through a game in Sunday School, a Bible verse was announced by its book, chapter, and verse number (eg., "*Salmos* (Psalms) 27:4"), and the first person to find the verse in the Bible, stand up, and publicly read it was triumphant. This motivated me to read the Bible in Spanish and design a strategy to memorize the sequence of books. I devised a code, such that the book of Genesis was assigned 1G, Exodus was 2E, and so on, in which the number was the order in which the book appeared, while the letter was the first letter in the name. I memorized this sequenced code, and while this was not a perfect system, it allowed me to play with numbers and letters and memorize the books in sequence.

Photo courtesy of Cynthia Anhalt.

This was an early memory of discovering how numbers play a role in developing my own schema with a purpose. While I remember this competition fondly, I never stopped to reflect on how children who did not excel felt during the game. I regret not showing my secret code to my peers, however, I am grateful that these experiences helped me to develop my literacy in the Spanish language.

My father and me.

## Early Education

My first address in the U.S. was 197 First Street, Nogales, Arizona, 85621, and this is where I observed my siblings do homework in their elementary years, and they tell me that I used to pretend to do homework alongside them by making tiny symbols on paper pretending to write the numbers and letters. I remember clearly looking forward to watching *Sesame Street* on television daily, and I looked forward to it because I could watch this show while everyone was in school except for my mother and me. I believe that I learned English as I counted numbers with Count von Count and sang with Maria and Big Bird.

As an adult, I learned that *Sesame Street*, produced in 1969 by a non-profit organization, specifically aimed to target underprivileged pre-school age children. *Sesame Street*, coinciding with Head Start, were products funded under the Economic Opportunity Act of 1964 signed by President LBJ, which aspired to provide access to education and economic opportunities long denied to low-income families. When I attended a Head Start Program in our community, I realized how much I had been looking forward to school because I saw my siblings attend school. I felt like I belonged in school.

The summer after first grade, I asked my teacher, Ms. Laz, for all the discarded mathematics workbooks with unused pages, and she gladly cleaned up and gave me the leftover books. I remember selling the workbooks to the neighborhood kids for 25 cents each, which were required for my tuition-free summer class in the front yard of our house. My parents were supportive, and got me a standing chalkboard that flipped over when I filled one side. I believe this experience convinced me that I wanted to teach mathematics.

While in second grade, I have memories of sitting in the back seat of our car watching moving numbers on the gas pump as fuel was going into our car at $0.36 and nine-tenths per gallon. It bothered me that I could not understand what the nine-tenths meant after

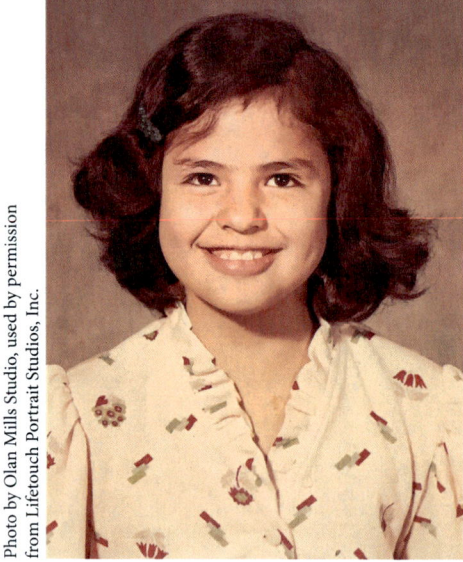

Third grade.

the 0.36 because after all, our money system only had up to the hundredths place value. I remember secretly playing a game of predicting how much the gallons would be after asking my dad how much money he was going to spend. I found the gas pump not easy to understand because the money amount and the gallon amount were changing at different rates.

While we attended school, my mother worked as a tailor making wedding dresses in Nogales. My father worked for Anamax Mining Company in Sahuarita, Arizona. One evening my father announced we were moving to Tucson, Arizona to the east side by Davis Monthan Air Force Base. This brought a big change to the family from a largely Latinx to mostly a white community.

At my new school, my third-grade teacher asked the class which fraction was greater, one-third or one-fourth, and I was the only one in the class who knew that one-third was greater. The other students thought that one-fourth was greater because 4 is greater than 3. The teacher asked me to explain to the class how I knew that one-third was greater. This was my opportunity to go to the board, grab a piece of chalk, draw pictures, write fractions, and explain my reasoning to my heart's content. This experience was exhilarating.

In sixth grade, my teacher Mr. Garbini, celebrated my learning like no other teacher

High school.

had ever done, and he made comments to me about how much mathematics I knew, which helped me gain more confidence. By eighth grade, when I was in algebra class, Ms. Stinson was the first teacher who used manipulatives to show concepts, such as two-color coins for positive and negative integers, and arbitrary lengths of small rods for variables. These visual representations of mathematical concepts made sense to me. By the end of junior high, I was the only student with a 4.0 grade point average, and I remember my friends' parents coming up to congratulate me after the awards ceremony.

In my senior year, I had an amazing calculus teacher, Mr. Dorsey, and he consistently lifted my spirits by periodically reminding me that I was the only Latina student in calculus in the whole school. This made me feel special somehow to know that he kept an eye on me so that I would succeed. He encouraged our small class of 12 to study together outside of class, so we did. A group of six of us would get large butcher paper from Mr. Dorsey on Fridays so that we could study together on the weekends and use the paper as our "white board." We were competitive but also collaborative with each other so much that we cared about each other's grades.

Outside of academics, I played the clarinet in the school marching band following my sister's example. I also had the privilege of competing in varsity sports, volleyball and tennis, and I especially excelled in tennis competing in the state championships for three years. My father inspired both my sister and me to play tennis, and each Christmas I got a new tennis racquet to prepare for the spring season. My father took me to a local private club for lessons every so often, and I felt privileged to be doing so with the professional instructor who was a former college and professional player. This must have cost my father a small fortune, but he wanted me to feel like I fit in. I believe that the competitive spirit with which I played tennis influenced my approach to all challenges. To this day, I enjoy playing tennis with friends of more than 20 years.

During my years in high school, and my siblings' years in college, my father decided to get his real estate license and became a realtor and a broker. It seems as though my family spent countless hours studying except for my mother, who would take an occasional English class for adults; she was the one who made sure we always had warm meals, clean clothes, supplies for school, and homemade Halloween costumes, always selflessly giving to the family.

## College

During my college years, I earned extra money by tutoring my friends in mathematics and found that I enjoyed teaching, which reinforced my childhood desire to teach mathematics. I enrolled at the University of Arizona following my siblings' footsteps as a first-generation college student on an academic scholarship. I do not remember ever meeting with an academic advisor except in my senior year when I decided I wanted to be a teacher. I had taken core mathematics courses (calculus 1 and 2, vector calculus, linear algebra, differential equations). When reviewing my transcripts, an advisor told me that I could combine the mathematics courses and the courses I had taken in chemistry, biology, environmental science, and physics, for an interdisciplinary degree and surpass the requirements in education. Teaching mathematics consumed my thoughts, and a few years into my career, I was invited to do teacher professional development in various school districts. Following these events, I began conducting peer professional development workshops in mathematics education for teachers. After that experience, I knew that I needed to pursue graduate studies.

## Graduate School

As I entered graduate school at the University of Arizona, I was a non-traditional student. I had taught public school for 10 years, had been married just as long, and had two children. In graduate school I renewed my interest in mathematics and mathematics teaching and learning. My advisor, Maria Fernandez, was my role model in that she was the only Latinx female faculty member in mathematics education, and we developed a life-long friendship beyond my graduate studies. She inspired me to become deeply engaged in the research literature in mathematics education, which consisted of research in teaching and learning of particular content areas in addition to understanding social issues within K–16 mathematics education. It is through experiential project activities, discussions, and analysis of published research that I began to understand academia in the field of mathematics education. Transition to Algebra was my first project in which I co-constructed and co-delivered professional development alongside Maria Fernandez for high school teachers. This project provided me an opportunity to study secondary teachers' metacognitive mathematical knowledge for teaching particular algebraic topics, and my first article publication resulted from this project.

My workload during graduate school became overwhelming as I was taking courses, reading research, and conducting research in mathematics education with the College of Medicine and College of Nursing at the University of Arizona. Our research team consisted of a medical doctor, a neuropsychologist, a pediatric nurse research scientist, and me, a doctoral student in mathematics education. Our study analyzed cognitive declines in children who were undergoing treatment for acute lymphoblastic leukemia, a common childhood cancer whose survival rate went from 33% to 80% within 30 years of medical advances, and thus the surviving children were growing up and becoming adults. Autonomously, I developed curriculum for teaching mathematics in a hospital setting for 7–15 year old patients who came for medical treatments for their cancer. They spent one hour per week with me doing mathematics activities prior to their chemotherapy treatments; now I reflect back and think what a torturous research study we conducted! From this project, I co-authored a research article, "Mathematics intervention for prevention of neurocognitive deficits in childhood leukemia," in the *Pediatric Blood and Cancer* journal, my first and only article in a medical journal. I was grateful for this opportunity to work in a medical setting, and I elaborate on this experience because I learned of the depth and breadth of mathematics education research, especially in interdisciplinary contexts.

Although I had unusual research projects during my graduate school, I pursued a dissertation research study focused on the development of mathematical knowledge in mathematical representation that teachers develop that is specific to the work of teaching. I was fortunate to receive support from my advisor, committee members, and my husband, who encouraged me to follow my academic dreams.

## Family Life

Coming from a small town in Wisconsin, my future husband, Dennis, came to the University of Arizona for graduate school in electrical engineering, and after we met, he decided to stay in Tucson, AZ. We were married in 1988 and have two amazing children. Our oldest, Ashley, was born in 1991 and Brandon in 1992. Through the years, it was interesting to see them develop their identities. When school assignments called for cultural integration, they would ask questions about their Mexican and German backgrounds, trying to make sense of the contrast between the two cultures.

Between 1998 and 2005, we traveled annually with five families to Punta Chueca and El Desemboque on the mainland coast of the Gulf of California in Sonora, Mexico to camp on the land of the Seri Indigenous people. We developed friendships with the Seri families

The family in 1999.

and traded our essential camping equipment, food, clothing, and bicycles among other things for their hand-woven baskets made of yucca plants. Influenced by these experiences, Ashley and Brandon wrote about them in school assignments. In 2013, we traveled to Yucatan, Mexico, where we immersed ourselves in the Mexican culture and had the privilege of visiting Chichén Itzá, an ancient Mayan ruin. Dennis and I wanted our children to grow up knowing their Mexican heritage and roots.

In our current lives, Dennis works for Texas Instruments as an engineer and manager of several teams of engineers across the world including the U.S., India, Mexico, China, Thailand, Malaysia, Taiwan, and the Philippines, and I am an associate research professor of mathematics education at the University of Arizona in the Department of Mathematics. Our daughter, Ashley, attended the University of Arizona earning dual bachelor's degrees, in mathematics and in systems engineering, and earned a master's degree in systems and industrial engineering from the University of Pittsburgh. Our son, Brandon, earned a

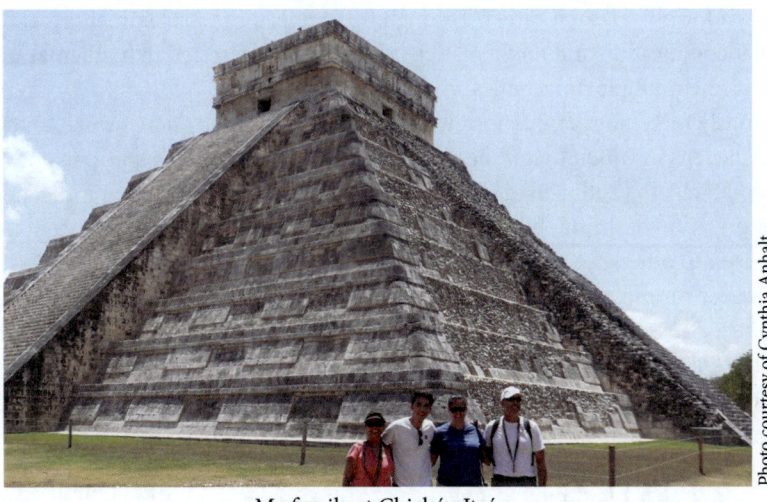

My family at Chichén Itzá.

Photo courtesy of Cynthia Anhalt.

My family at home.

bachelor's degree in physiology from the University of Arizona, and earned his medical degree from the Medical College of Wisconsin. I am proud of our accomplishments as a family, and there has been much sacrifice in the time we have spent pursuing our achievements, and in recognizing this, we appreciate the time that we spend together.

## Final Thoughts

My path to becoming a faculty member was non-traditional. Upon graduating with my PhD, I became the director of the Center for the Mathematics Education of Latino/as (CEMELA), a National Science Foundation Center for Learning and Teaching in collaboration with three other universities. After one year of administrative service as director, I became a postdoctoral fellow since the project goals were a perfect match for my research interests in the mathematical preparation of teachers of underrepresented Latinx/Hispanic student populations in mathematics. CEMELA, under the direction of Marta Civil, served as a catalyst for my research career. Through this project, I met prominent researchers in mathematics education and mathematicians genuinely interested in mathematics education across the multiple universities.

I curiously remember discussions with my family around a quote, which to this day influences the lens in which I view the world, "*El respeto al derecho ajeno es la paz,*" spoken by Benito Juárez, the 26th president of Mexico and the first president of Indigenous origin. My interpretation of the quote, "respect for the rights of others is peace" meant that I should listen to others, appreciate their stories, and understand our differences. I raise this point because for most of my life, I presumed that others would treat me as I treat them, with respect. This presumption was not always the case in my academic experience. In my position as an Assistant Research Professor and director of the Secondary Mathematics Education Program (SMEP), tenured faculty members in my department raised questions about my junior faculty status toward gaining tenure. I suspected that the questions raised were regarding my scholarship mainly due to the decreased workload percentage in research, yet my research publication record proved exceptional, and I became an Associate

Research Professor. I continue to serve on department and college-wide committees, for example, I served on the UArizona College of Science dean search committee.

I continue as the director of SMEP and participating in grant-funded projects. As principal investigator (PI) of the Arizona Noyce project funded by the National Science Foundation (NSF), my focus was on preparing highly qualified mathematics teachers for diverse student populations. I was PI of the *Mathematical Modeling in the Middle Grades* project funded by the Arizona Board of Regents to deepen and broaden teachers' knowledge of mathematical modeling for teaching. I led professional development in the *Mathematical Modeling in Cultural and Community Contexts* project funded by the NSF, and created curriculum materials in mathematical modeling for secondary teacher preparation for the project, *Mathematics of Doing, Understanding, Learning and Educating for Secondary Schools*, funded by the NSF. My research publications focus on mathematics teacher education with emphases on mathematical modeling and development of teaching practices for inclusion, equity, and social justice. I have been privileged to participate in national and international mathematics education conferences presenting my research.

While on sabbatical, I had the privilege of co-leading with Rachel Levy, from the Mathematical Association of America, the 2019 Critical Issues in Mathematics Education Workshop at the Mathematical Sciences Research Institute, which focused on mathematical modeling in K–16 education. This workshop has been a productive setting for developing partnerships among mathematics educators and mathematicians interested in improving K–16 education. One key underpinning of my career has been to build lifelong collaborations that aim to move mathematics education research forward, and I have had the honor to collaborate with colleagues, Ricardo Cortez, Maria Fernandez, Julia Aguirre, Rochelle Gutierrez, Sylvia Celedon-Pattichis, Sandra Crespo, Anthony Fernandes, Marta

With Rachel Levy.

Photo courtesy of David Eisenbud.

Photo courtesy of Cynthia Anhalt.

With Ricardo Cortez and Rochelle Gutiérrez.

Civil, and others. I am grateful that I was able to follow my passion and pursue a career in mathematics education.

## Advice

Photo courtesy of Eric Landwehr.

Dr. Cynthia O. Anhalt.

My advice to students is to pursue their dreams and passion, and use their knowledge of mathematics as a foundation for a fulfilling career. The career may be in mathematics, teaching mathematics, actuarial work, data science, and graduate opportunities in mathematics. Sharing your story with a mentor can be powerful. I share my story with students, and I appreciate their reactions and their enthusiasm for sharing their stories with me, as I believe that these interactions help build community and long-term relationships. My hope is to promote access and options, especially for Latinx students when they share their hopes and dreams.

# 17

# Dr. Omayra Ortega

## Why Math

I have always been drawn to beauty and order, two concepts central to mathematics. I grew up in Far Rockaway, Queens, New York—a wonderful land on the south shore of Long Island (right near the beach). The ocean has always been a tranquil escape for me, and I've always felt a connection to my ancestry through the shared experience of the ocean. As a child in the early 1980s, I reluctantly spent many hours (on both weekends and weekdays) with my parents at church. Rather than engage with the spoken gospel, I would admire

Dr. Omayra Ortega

Illustration created by Ana Valle.

the interior of our church, St. Mary's Star of the Sea, and take in the physical gospel that was presented before me. During these long masses, I would get lost in the lights and the architecture. I loved the symmetry and the grandeur of the altar and how that stately mirror-like design was continued throughout the entire building. I was especially hypnotized by the stained glass of the stations of the cross and the stylized geometry of the figures contained in each of the colorful panes of glass. Each image made more beautiful by the mid-morning sunlight. One of my earliest memories is waiting outside of the church after my brother's youth bible study group, *Las Jornadistas*, got out. All of the young people were happily chatting and milling about before walking home. My brother and I got a rude surprise when we realized that his bicycle—our only mode of transportation at the time—had been stolen while we were in church. To this day we still marvel that someone would steal a bicycle with a baby seat attached from a church! It takes all kinds.

My homeland, New York City, is a melting pot of people and cultures, and Rockaway, Queens had a similar mixture of cultures, but on a smaller scale. Rockaway Beach is a long peninsula on the south side of Long Island with a long swath of sandy coast facing the Atlantic Ocean. Rockaway Beach is defined by Breezy Point on the western border and Far Rockaway (where I was born) on the eastern border. All types of people are drawn to this area because of the beach and relative proximity to New York City, but the neighborhoods are somewhat segregated. Breezy Point tended to be more well-to-do and working class people of European descent. Many of my friends from this area were Irish Catholics.

My grandmother, in a *pollera*, with my mother and two uncles (everyone is wearing traditional Panamanian dress).

Breezy Point, Neponsit, Belle Harbor, Rockaway Park and Rockaway Beach are home to many of the firefighters of Irish descent who responded to and unfortunately perished in the 9-11 attacks on the Twin Towers (may they rest in peace). As you move further east on the Rockaway Peninsula into Arverne, Edgemere, Bayswater, and Far Rockaway, you would find the homes of the Black and Latinx Rockaway residents. The easternmost border of Far Rockaway was (and is) defined by a large orthodox Jewish community, beyond which lay the suburbs of Nassau County, what many people call the start of "Long Island" (though all of Brooklyn, Queens, Nassau and Suffolk define Long Island).

## My Familia

If your family is anything like my family, every family function is the same. It didn't matter if it was a baby shower, kid's birthday party, *quinceañera*, wedding, graduation party, or a funeral:

(1) There would be food.
(2) There would be drinks.
(3) There would be music.
(4) There would be dancing.

Omayra as a little girl.

There would be your entire multi-generational overflowing family. The adults would dance and drink while the kids would go off with their snacks (popsicles, *empanadas*, and/or birthday cake) to play Hide-n-Seek, tag, or Nintendo (I am classically trained on the original NES, if you're wondering). My family is almost exclusively Panamanian, it's mainly my generation that has started "mixing" and marrying non-Panamanians (*ay Dios mio!*). Panama is a beautiful tropical isthmus with the Atlantic and Pacific Oceans on either side. We love our *sancocho* (soup), *frituras, tamales, arroz con pollo, arroz con habichuelas*, and most dear to our hearts, *PLATANOS* (plantains). Both of my parents emigrated from Panama when they were young—my mother when she was 14, and my dad when he was 18—both looking for a better life. They found each other after moving to New York City at—you guessed it—a Panamanian party. They married shortly after meeting, moved to Brooklyn, and started a family *immediately*. My two brothers Oscar and Omar were born within one year of each other, more than a decade before me.

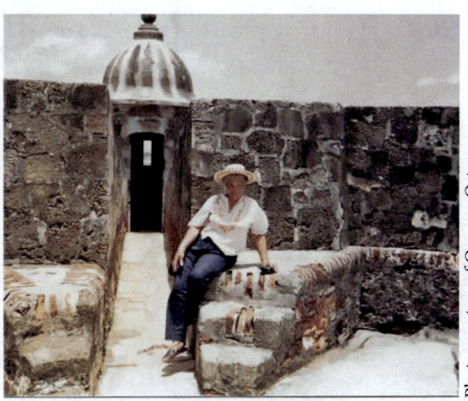

Omayra's Grandmother.

Photo courtesy of Omayra Ortega.

My father assures me that the family was not quite complete until I appeared, he says *"siempre queria una hembra,"*[1] and 12 years later, I made my appearance on this earthly plane. My mother assures me that all of her children were accidents, something I don't doubt considering the lack of sex education in Panama City and New York in the 1950s. By the time I was born our family had moved into a house in Far Rockaway, Queens, New York, where we often held family gatherings in our basement, in our backyard, or at the beach.

Everyone in our large extended family hosted family gatherings. My grandmother, Nana, was the matriarch of the family and frequently hosted us all in her home in Cambria Heights (Queens). My grandmother meant the world to me and I probably spent half of my childhood in her home.

## Early Education Years

I remember after starting elementary school she asked me why I let my friends in school call me *Oh-My-Rah*, and I remember her face twisting up to say the letter 'R' in her most exaggerated American accent. I didn't know how to ask people to say my name correctly. I didn't know how to take up space yet, and I'm not sure that I've fully learned that skill even now, even though it means the world to me when people try to say my name properly. The very first time a teacher said my name correctly I was in college. I honestly almost cried when my ethnomusicology professor, Katherine Hagedorn, said my name correctly without any instructions from me. It has happened again, but it is still rare. It's especially hurtful when people don't even try to get my name right and insist on butchering it like it's not my own name, and they somehow know how to pronounce it better than I do. FYI—single 'R's in Spanish sound just like a soft American 'R,' so you should not roll them (like a double 'R' in Spanish), and your American mouths should not have a difficult time executing them—it's in your tongue's lexicon! You can do it! (*Si se puede!*)

Another important host for our family gatherings was my *Tia Marta*. This *tia*[2] made some of the most delicious food in our family. She was always ready with a FEAST. I should add that this *tia* was not a blood relative, but had been partially raised by my grandmother. I have a very large family, in part, because if a family friend spent enough

---

[1] Translating to "I always wanted a little girl."

[2] Aunt.

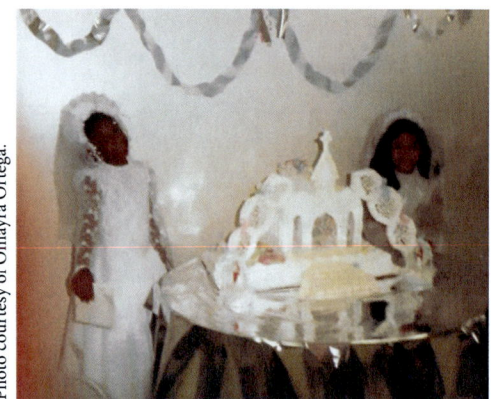

Omayra and cousin at their first communion celebration.

time with us and had integrated themselves into several generations of our family, then they became family themselves. I think it is age that designates whether the person becomes an aunt/uncle or a cousin, but once they are part of the family, so are all of their descendants. Eventually they become blood. We also never differentiate between first cousins, second cousins… you're just a cousin, and that is that. My *tia* had a daughter who was a constant in my early life. Our mothers were pregnant around the same time and my cousin was born 11 days ahead of me. We lived in the same town, and were the only girl-children in our generation, so we went to the same pre-school, we had frequent play dates, we would sit together at church, we even had a joint first communion celebration.

She was my very first friend and my BEST friend. When we started kindergarten at P.S. 183 in Rockaway Beach, Queens, I assumed that she would continue to be in my class

Dr. Ortega's great-grandfather, Enrique Alonso Duesbury (right), and his brothers. McDonald Duesbury (great-grand uncle, middle) sponsored Dr. Ortega's grandmother's and children's immigration to the U.S.

since we'd been inseparable since day one (literally). I knew that I was going to be enrolled in the Astor Program for Gifted Children because of some tests I had taken, but my five-year old mind didn't really understand the implications of this. I was heartbroken when we were separated and she wasn't in my class.

At the time, the New York Public School System was tracked. Upon entering, students were sorted into one of three tracks: either the Astor gifted track, Regular Education, or Special Education. I felt honored to be in a class with other "gifted" children—I loved the new things that we were learning—but even then, it felt wrong to be separated from my cousin who, up until that point, I had spent my entire life with. I missed the comfort of her company and I didn't know any of the other pupils in my class.

There was one Astor class for each grade and the racial breakdown of the students between the three tracks

was stark, even to a five-year old. Most
of the students at P.S. 183 were Black or
Latinx. These two subgroups defined
almost 100% of the Special Education
and Regular Education classes at that
school. However, out of the 35 pupils in
my kindergarten Astor class, only seven
of us were Black or Latinx; the remainder
were white. The contrast was so stark to
my young eyes, and I still remember the
full names of the other six. These num-

Omayra at Milton Academy.

bers grew *slightly* each year, but never enough to tip the balance. The segregation that I
observed on the Rockaway Peninsula seemed to have been maintained in my elementary
school classrooms and would be a theme throughout my life.

When I was around 12 years old I was nominated to apply to Prep for Prep 9 by my
teachers at J.H.S. 180. Prep for Prep 9 was a program based in New York City that found
talented youth from around the city and prepared them to attend prestigious boarding
schools. This program changed the trajectory of my life. At a young age I decided to leave
my parents' home in Far Rockaway and move to a suburb of Boston for schooling. We all
saw this as an opportunity that I could not pass up, and I was ready to leave my parents'
conservative home. While I loved spending time with my friends and family at church, the
idea of "a woman's place" was one that I could not fall in line with, so I was ready to start a
new chapter in my life.

Attending Milton Academy in the early 1990s was an academic blessing and a welcome
escape from my conservative home in New York. I knew that this was a great opportunity for me, but it also took me away from my family and my culture. Prep for Prep 9 had
prepared me for the academic rigors, but I was not ready for the extreme wealth of my
classmates and the new social expectations. I thoroughly enjoyed my four years of high
school—I made great friends, did well in my classes, and held several leadership roles—
but issues of race and class were always in the back of mind.

## Undergraduate Education and Introduction to Mathematics

I came to the field of mathematical epidemiology in a round-about way. When my dreams
of becoming a medical doctor were dashed by my lack of success in my first-year general
chemistry course, I focused my attention on my double major in mathematics and music.
I loved my time at Pomona College. In retrospect, I can say that moving from the east
coast to the west coast was one of the best decisions that I made in my young life. I can
say this now because I still live in the beautiful state of California, but, during that first
year, I wasn't so sure. The social aspects of college life were most enticing to me, and, after
my first semester, I was put on academic probation. Despite "trying to do better," I was
suspended for one year after my second semester at Pomona College and had to attend
another school for one year and earn a B or better in all of my classes there. I wasn't ready
to go back to the east coast (read: go back to church), so I went to live with my *Tia Nory*
and *Tio Calin* in Milpitas, San Jose, California and attended DeAnza Junior College in

Cupertino. It was nice living with my maternal uncle's family. I was able to live with two cousins close to my age and experience a portion of my "coming of age" years with my family. It was a challenging year where I worked full time and took more than a full load of classes at DeAnza, but this year gave me time to reflect on exactly what I wanted out of life and out of college. Gap years were not a thing when I went to school and certainly not a thing for people like me with working-class immigrant parents, but I understand how taking a gap year can give you extra time for reflection and for establishing life goals. While I enjoyed DeAnza to the fullest by joining every musical ensemble and taking most of the music courses offered on top of taking the required transfer courses for my majors at Pomona, I knew that I didn't want to stay there any longer than I needed to. I knew that I wanted to return to Pomona College to complete my degree and I wanted to thrive there.

After that year, I was able to return to Pomona College and continue my degrees in pure mathematics and music performance. It wasn't all roses, but I was successful and definitely improved over my performance during my first year. During my junior year I was encouraged to apply for a summer research experience (REU). These REUs were new at the time, but are pretty common now. I spent the summer between my junior year and my senior year attending the prestigious Mathematical and Theoretical Biology Institute (MTBI) which, at that time, was held at Cornell University. This program was formative for me in that I conducted research for the very first time, met many of the people who are my collaborators and colleagues, and chose my field of study: mathematical epidemiology. I also learned what it is to work, very hard, on a collaborative project. I was able to revive my deferred dream of working in a medical field. I learned that through mathematics, I could still be a healer by working in mathematical epidemiology and public health. In this program, I worked the hardest that I ever had. We packed SO MUCH into the short eight weeks that we had together. I acquired the basic skills in math modeling and differential equations that I continue to use to this day. Those sleepless nights also trained me for when deadlines approach faster than you expect.

One of the most important memories I have from this program was when Dr. Colette Patt came and gave a presentation on the state of mathematics PhDs. I remember hearing that less than 2% of the PhDs awarded in mathematics in the previous year were award-

Photo courtesy of Omayra Ortega.

College graduation.

ed to people of color and less than 1% went to women of color. Those statistics made me very angry but also motivated me to pursue a PhD in applied mathematics. Those statistics haven't changed much, so they continue to motivate the work that I do, to this day.

The year before I started my PhD program, I participated in a summer program at Spelman College that set me up for success during my first year in grad school. I am very thankful to have participated in the Enriching Diversity in Graduate Education (EDGE) Program for Women. Through that program I got a more realistic perspective of the academic and social challenges that I would face during my first year, and EDGE gave me a network of supportive sisters and allies that would help me to succeed in the early years and throughout my career.

# Graduate School

I didn't get into grad school the first time I applied, but I did get in the second time (*gracias a Dios*). On the recommendation of my mentor, Carlos Castillo-Chavez, I went to the University of Iowa to study mathematical modeling of infectious diseases under Herbert Hethcote, one of the founders of the field. I had never been to the Midwest, so I experienced yet another period of culture shock and then acclimation. If you've never experienced a midwest winter, you are a blessed individual. The University of Iowa had recently been awarded a big Graduate Assistance in Areas of National Need (GAANN) grant from the National Science Foundation so I was very lucky to receive a fellowship to support my doctoral studies. This GAANN grant built off of the previous efforts of professors at Iowa to bring more students of color to the lush, verdant corn and soy fields of Iowa. Thanks to the efforts of professors like Gene Madison, Phil Kutzko, Yi Li, Juan Gatica, Richard Baker, and David Manderscheid, the University of Iowa Department of Mathematics won the Presidential Award for Excellence in Science, Mathematics and Engineering Mentoring in 2004. The math department at Iowa was a really warm and supportive community. Everyone helped each other if we were having trouble in classes, if we needed a ride to one of the (distant) airports, or if we needed help moving into a new apartment. I remember being so worried about my partial differential equation (PDE) comprehensive exam. I knew that I was doing fine in the class itself, but I wasn't sure that I could complete those same types of problems in a classroom space not of my choosing and in a specified time frame. I normally worked on homework problems at all times in all spaces. I could be in my office on campus, working at a nearby coffee shop, or at home asleep, and the solution would come to me. The classroom really wasn't where I did my best work. Several older graduate students made a point to share their study binders from previous PDE comprehensive exams and helped me to study. I am so thankful for their generosity because they helped me, not only to prepare for my exams, but allowed me to feel relaxed enough to perform at my best on the day of the exam (BTW—I passed with flying colors!).

We were also there to celebrate each other if someone in the math department had a child, passed a comprehensive exam, or successfully defended their thesis. I remember coming together frequently at Phil Kutzko's home, always over food and drinks. I loved the constant celebrations and get-togethers within the math department, as they reminded me of the way we would celebrate family at home in New York. Most of the math department were regulars at one bar and grill on Tuesday evenings where we took over many tables with our textbooks, pint glasses, and wing baskets on 'Tuesday Wings Night,' and we ended the week at another bar where we would lick our wounds and recap the week during 'Friday After Class'. I really don't think that I could have completed a PhD anywhere else. I have lots of colleagues in mathematics now, many who did not go to Iowa (*no me digas?!*) and, when I hear stories about their grad school experiences, I realize how lucky I was. Nowhere else would have supported me in the way that Iowa did. I felt nurtured both academically and socially inside the department, at the coffee shops, and even at the bars.

I spent the last two years of my PhD in an all-but-dissertation instructor position at Arizona State University. Those last two years teaching full time while trying to complete my PhD were arduous. I would not wish that experience on anyone, and I highly dis-

Omayra and her mentor Dr. Carlos Castillo-Chavez.

courage anyone from attempting to start a new position before completing their PhD. Even though it started off rough, I am thankful to have had this opportunity because it led to a nine-year career at ASU. There were many moments when I thought I should just quit and be happy with a master's degree in mathematics and another in public health, but I am glad that I persevered. My parents felt that I might as well finish since I had come so far, but they would support me if I decided to quit. Truthfully, they had no idea why I was still in college anyway, they hadn't quite grasped the idea of graduate school (there's more college after college?). I truly appreciated the professors from Iowa and Carlos Castillo-Chavez, who was at ASU at that time, who would occasionally check up on my progress at conferences, through email, and text messages. Without these ever constant, "How's the thesis going?" questions, I might have given up. I felt the weight of what I owed to these mentors and the breadth of what they knew I could achieve. I am thankful that these mentors did not give up on me when it seemed like I was dragging my feet through the last phase of my thesis work.

Because each successive step of my schooling moved me further and further away from my family, it took a lot of effort to maintain ties with my family and my culture. In some stages of my life, I was not successful or even motivated to maintain these ties. I am glad that I am able to reflect on this phenomenon now that I am an adult. In my day-to-day life, I find myself thinking more and more about how I can "decolonize my mind" for myself, for my students, and for my family. With each step in my education, I lost my connection to the Spanish language. We spoke Spanish in the household when I was little, but as soon as I started kindergarten, my parents only spoke to me in English. Even to this day, I have to work to get my parents to speak to me in Spanish. Even if I start the conversation in Spanish they naturally revert to English, to which I reply either "*que?!*" or "*como?!*" to get them to go back to their first language. Through my two brothers, I have four nieces and nephews and none of them even have a basic knowledge of Spanish. I see that part of our heritage getting watered down with each successive generation, and it makes me sad. Though I recognize both English and Spanish as "the colonizer's language," I'm trying very hard to keep my Spanish language proficiency up.

## Applied Mathematics and Public Health

I grew up on Rockaway Beach and spent almost every summer day at the beach. I could play all day in the dancing ebb and flow of the waves. It's one thing to see the dynamics of an ocean wave from the shore, but it can be a transcendent experience to be carried by a wave and to dance along with it. I love that we can use partial differential equations to describe the dynamics of the ocean's waves. The poetry of mathematics is all around us, even when we are not aware. Though my love for differential equations stems from the modeling of infectious disease, I still marvel at how differential equations are applicable to so many different natural phenomena.

Mathematical modeling allows me to describe the world with mathematics. I like to think that math is a universal language and, through modeling, we can share poetry about the world that we live in. The research that I've conducted in mathematical epidemiology first started that summer at MTBI. I worked with two other students modeling the evolution of drug-resistance in the yeast *candida Albicans*, using a coupled logistic equation model to describe competition between two strains of *c. Albicans* in a human host and the effect of using an antifungal agent to try to control their growth. From then I always focused my work on emerging infectious diseases, new vaccines, or tropical diseases.

My research in mathematical epidemiology is grounded in the public health problems of today. Using mathematical modeling and the theoretical core of mathematics as tools, I am able to better understand and describe emergent health problems such as HIV, HPV, rotavirus, malaria, polio, and TB. My contributions to science and society improve the health community's understanding of infectious diseases and inform policy makers on how these diseases can best be controlled at the population level. In order to be well prepared to contribute to this field of mathematical epidemiology, I simultaneously study applied mathematics, statistics, epidemiology, and public health.

Currently, I work with a fantastic set of collaborators on mathematical models for coronavirus where we evaluate different isolation strategies and their cost-effectiveness. We also developed geo-spatial models for the spread of malaria which takes into account immigration, seasonal migration (i.e., seasonal workers), and tourism between Botswana and its neighboring countries. I first met these collaborators, who are a diverse group of women, at a workshop organized by the Association for Women in Mathematics at the Institute for Pure and Applied Mathematics, under the name, "Women in Math Biology" (WiMB). I am incredibly thankful that I could participate in this workshop because it helped restart my research program, which had been focused on undergraduate research only for about five years. I love developing research skills in undergraduates and I still maintain my Mathematical Epidemiology Research Group, comprised of all undergraduates; I also work with the Rocky Mountain Sustainability and Science Network every summer, but it is nice to focus on my own work and publications sometimes. I am also working on a project with colleagues from Sonoma State University (SSU) and other experts from other institutions trying to identify both institutional and implicit sources of bias in STEM, starting with the math department at SSU as the pilot study. This work lies at intersection of my service work and my research, so this is an exciting new venture for me—one, which I believe, could have only happened at my current institution. My experience at Sonoma State University, though still new, has been both refreshing and eye-opening after teaching at three other institutions of higher education.

## The Search for Balance and Advice

It has been very nice to finally understand that I need a balance of teaching, research, and scholarship to be happy. If I focus on just one aspect alone, I feel that something is missing. Being at Sonoma State University has allowed me to continue my devotion to my students through teaching and research, to continue my scholarship and publications in the modeling of infectious disease, and to continue my service, not only within my own university and department, but also nationally through my service work with the National Association of Mathematicians, the Association for Women in Mathematics, the Society for the Advancement of Chicanos and Native Americans in Science, and the Mathematical Association of America. If there is one piece of advice that I can give to folks about to embark on a career in mathematics—one piece of advice I wish that someone had given me—is that you should thoughtfully choose the institutions where you will work, study, and devote large swaths of time, based on your interests and the type of life balance that you would like to have. Don't be afraid of changing institutions if the institution that you start your career at is not a good fit for you. It's better to have a delayed start than to finish at an institution that doesn't work for you or, worse, *not* finish because you were at an institution that didn't work for you. I didn't always make the best choices initially, but that is really how life experience works. I am, and always have been, on the right path—my own path. Hold tight to your culture and your life goals. Use them as your North Star and Southern Cross to navigate the inevitable ebbs and floods that will cross your path.

# 18

# Dr. José A. Perea

## My Life through Mathematics

The opportunity to write this chapter has been a breath of fresh air amid very tumultuous times. It is my sincere hope that when we look back to the year 2020, it will be as a watershed moment where a large majority agreed that racial injustice, xenophobic sentiments and less than competent leaders cannot be tolerated in society. Something that has come up again and again during this reflection is the idea that every perceived personal success has had many people behind it. Mentors who, with small acts of kindness, have a huge impact. I will try to describe a bit of my journey as a vehicle to articulate these thoughts.

Dr. José A. Perea

Illustration created by Ana Valle.

**Early life**. I was born in 1984 in Santiago de Cali, the third most populous city in Colombia, to Adiela Benítez and José Lúcio Perea. Geographically speaking, Cali is nestled between the central and western Andean mountain ranges; it has a beautiful year-round tropical climate, and it is a rich melting pot of indigenous and Afro-Latin cultures. People refer to it affectionately as *La Sucursal del Cielo* (Heaven's branch office) or *La Capital Mundial de la Salsa* (The world's salsa capital). The Afro part of Cali can be traced back to several migrations, particularly from the Colombian Pacific. This predominantly Black part of the country emerged from early slave settlements in territories like Chocó—a region historically plagued by poverty and government neglect. My dad and my maternal grandma, Rosa Lia Mosquera, were born in the neighboring river towns of Santa Rita and Santa Ana, Chocó. I don't think they knew each other while living there (age difference), but they both emigrated to Cali in search of a better life. They came with nothing; not even a high school diploma, but they both built a life for themselves as immigrants. My grandma worked very hard to singlehandedly raise my mom, and my dad put himself through school—almost finishing a university degree in accounting—while building a small tailoring business to support my family.

I credit my dad with breaking a very resilient cycle of poverty in his side of the family. When I was around fifteen years old, he took me on a trip to his childhood home in Santa

Photo courtesy of José Perea.

My family (February, 1999). L to R: My sister, my mom, my dad, my grandma, and me.

Rita. I remember the two of us traveling by bus from Cali to Chocó, and then helping paddle a small boat for the better part of a day—a four-person canoe—along the Iro river. We finally made it to Santa Rita, where I found a beautiful town filled with warm and welcoming people, though with few paved roads and even fewer homes with electricity or running water. My dad's home was not one of those few. Going back to those memories always makes me think of the effect that initial conditions can have, the importance of pushing for equitable policies, and of our responsibility as mentors to not let said conditions get in the way of someone's work ethic and talents.

Growing up we were not middle class, but my parents always made sure we had everything we needed. My mom and my grandma would repeat mantras like "being poor is not an excuse for having dirty shoes or wrinkled shirts," which were part of a bigger theme. For example, in middle school I had horrible handwriting; my mom would sit with me to go through my notebooks, she would take out all the messy jumbled pages and make me rewrite them neatly. Today I have better handwriting thanks to her, and an appreciation for the benefits of putting effort into getting the small things right. I believe these were, in their way, an attempt to lessen our chances of being subjected to discrimination later on. My grandma had a line that went something like this: you'll be judged as a Black man first, and your actions will reflect on Black people in general; it will not be José did this or that. As unfair as it sounds, this was in all likelihood a result of her upbringing, interactions with other people, and personal journey. I've had similar feelings as an immigrant in the U.S.; that I would be judged as Colombian first (I've heard all the cocaine-related "jokes" and they are still annoying) and that my failures could reflect poorly on an entire country.

While these feelings are legitimate and, I imagine, shared by others, it is crazy to think that minorities—in addition to everything else—have to deal with the constant pressure of being "the" representative. I hope we can dispel these notions through honest conversations with our students, while we help build more diverse environments where people don't have to feel this way.

**High school years**. My love for math came almost at the end of my high school years at *Centro Educativo Industrial Luis Madina*. I had always been a good student with top grades, and learning was something I enjoyed, but I wouldn't say that math was something I was passionate about. Physics, on the other hand, blew my mind. When we learned Newtonian mechanics, it was incredible to me that one could understand and formalize the world with math. I was also very lucky to have an unconventional physics teacher, Prof. Jesús Rivera. Once he noticed I was doing well on the tests and reading ahead in the book, he told me to go to the school's library instead of coming to class; that I should study the material at my own pace and periodically talk to him as I made progress. This freedom to learn was empowering, and I was convinced that physics would be my major in college.

In Colombia, students declare their major when they apply to a university; they submit their materials to a specific program and are admitted according to a ranking of test results.

The test in question is called the ICFES exam.[1] It is a standardized test like the SAT, administered nationally to all high school seniors, and used by Colombian universities to determine admission. Needless to say, it is a big deal. The large public universities in Colombia are quite good, though there are few of them, and the ICFES scores are essentially their only admission criterion. Thus, for students who cannot afford private education, it is a high-stakes exam. This was certainly my situation, and my mom embarked us on a mission to make sure I got into *Universidad del Valle* (*Univalle*)—the third-largest public university in the country, located in Cali, and ranked among the top five nationally. She signed me up for a pre-ICFES course that met every Saturday. I was of course very reluctant—I thought I could study on my own—but I was pleasantly surprised with the unforeseen benefits.

The preparation course was structured to review all relevant material for the test by splitting the content into classes. Among them was obviously mathematics, but the course was entirely subverted by the teacher in charge. On the first day of classes, he said, It is very unlikely we'll be able to cram in a few Saturdays what you haven't learned in six years; let's learn something interesting instead." The teacher was an undergraduate student in mathematics at *Univalle*, and he proceeded to review several "math facts" from one viewpoint—Why is this true? We covered several topics in arithmetic, algebra, and geometry from a discovery/proof-based perspective, which was entirely alien to me. Until then, I had seen math as a formulaic and rote memorization exercise, which is unfortunately still very common. From then on, I was hooked.

The course ended, and the test came and went. In the end, I got the top ICFES score in my school, and the question became whether I would go for physics or mathematics in my application to *Univalle*. I didn't have a good framework to make this decision, so I went to

[1] The ICFES stands for *Instituto Colombiano para la Evaluación de la Educación* (ICFES).

Prof. Rivera for advice. He asked me, "What it is that you find interesting?" I responded, "Figuring out why things work, why they are true." Without missing a beat he said that I'd probably be happy doing math, and several years later I can report that he was 100% correct.

I went to *Univalle* for my bachelor's in mathematics between 2001 and 2006. As I mentioned before, the Colombian higher education system is geared towards students focusing on their major almost exclusively from day one, though other complementary courses are also included. By design, I ended up taking a lot of math classes: from number theory and set theory to abstract algebra, real and complex analysis, measure theory, functional analysis, differential geometry, and point-set/algebraic topology. Everything seemed super interesting. Along the way, I met several professors who inspired me to pursue mathematics, and selflessly helped me succeed. Among them, I want to highlight Dr. Doris Hinestroza, Dr. Gonzalo García, and Dr. José Raul Quintero. They all graduated from *Univalle* and went on to earn PhDs in mathematics from universities in the U.S.: Doris from the University of Cincinnati, Gonzalo from Cornell University, and Raul from the University of Maryland. Doris passed away in February of 2019; she was kind, warm, generous, a force to be reckoned with, and genuinely excited about doing and teaching mathematics. She is sorely missed.

During my third year in college, I took Real Analysis with Gonzalo. My friends and I used to study for tests by attempting to solve all problems in the book (in this case Rudin's *Principles of Mathematical Analysis*) for the units covered in the exam. Of course, there were problems we couldn't do, so we would go to office hours often, and talk to Gonzalo about everything math-related. At the end of the semester, I told him that I wanted to learn algebraic topology, and asked if he would be willing to advise me. He agreed and the very next semester we set up a reading course (Massey's *A Basic Course in Algebraic Topology*) and with his guidance, I was able to complete a laureate thesis titled *The Borsuk-Ulam theorem and its applications*. I thank him not only for his guidance but for the many opportunities he facilitated so that I could be successful. One of the first things he told me when we started working together was that he thought I was capable of getting a PhD overseas, but for that to happen I needed to speak English. At the time I could read it, but I certainly couldn't write/speak or understand it without subtitles. Gonzalo, through his contacts, found me a job as a substitute math teacher in a bilingual (English-Spanish) school, and with the salary, I was able to pay for a one-year intensive English course. When the time came for me to take the TOEFL, GRE and GRE-math exams, Gonzalo let me use his credit card (my family didn't have one) to pay the online registration fees. Of course, I paid him back later, but these were all things he didn't have to do, which were huge for me. I cannot thank him enough.

Looking back, I now think of these logistical challenges as examples of structural roadblocks where the socioeconomic background can limit access: did you go to a school where learning to speak a second language was possible/encouraged? Did your parents have resources available for extra-curricular/academic activities? Did you have access to people who knew the system and could give you timely and accurate advice? I was lucky to have people in my corner who helped me circumvent these roadblocks. The goal should be for all our students to feel that luck is also on their side.

Photo courtesy of José Perea.

College graduation (April 2006). Doris was the dean of Natural Sciences; I was super happy to get my diploma from her.

I graduated from *Univalle* in April of 2006 (Valedictorian) and started my PhD in mathematics at Stanford University in September of the same year. The application process for grad school started two years earlier, from the time Gonzalo suggested I learned English, to one afternoon in Doris' home with her, Raul and Gonzalo helping me prepare the application packages—at the time these documents (translated and notarized grades, recommendation letters, CV, essays, etc.) needed to be physically mailed. The three of them wrote recommendation letters on my behalf and helped make sure that the right documents were in the right envelopes for each university. Due to costs, I could only apply to five schools; all in the U.S. and across a wide range of academic rankings. I was denied admission in two, short-listed for one, and admitted into the other two.

## Stanford Years

The time I spent in California as a graduate student was filled with good experiences, both academic and social. I made a lot of good friends and learned a lot of mathematics. And even though I worked very hard, I vividly remember feeling like an impostor, that the admissions office had surely made a mistake when they let me in. There is actually a name for that, the *impostor syndrome*, and it must have been widespread enough at Stanford because there were several institutional initiatives to help combat it. For instance, I attended a university-sponsored workshop for coping mechanisms, in a packed auditorium, and there was also this very cool series called *The Resilience Project*, with videos[2] from top faculty—heroes of many—telling their stories of academic and professional failure. I had my share of those in grad school. I failed my first midterm ever within two months of starting (measure theory) and also failed the algebra qualifying exam. As painful as they were, those failures were useful. All the algebra I learned during the summer of that year, while studying to retake (and pass!) the qualifying exam, has helped me in my research to this day.

[2] *The Resilience Project* series is still available here: youtube.com/channel/UCWImwo-jhbas29JwQlslsLGg

Photo courtesy of José Perea.

PhD graduation (June 2011) with my parents.

Speaking of research, I started graduate school convinced that I wanted to focus on algebraic topology. This is the branch of mathematics concerned with spatial properties that are invariant under continuous deformations. For most of its history, algebraic topology has been regarded as a purely theoretical subject. Imagine then my surprise when, during one of my reading courses, I found out that topological ideas were actually being used to solve real problems in data analysis. Today, computational and applied topology, and in particular topological data analysis (TDA), are rapidly developing disciplines at the intersection of statistics, mathematics, and computation, with many students, vibrant conferences, and faculty in universities worldwide. In its inception, the main advances in TDA were spearheaded by Dr. John Harer and Dr. Herbert Edeslbrunner at Duke University, and by Dr. Gunnar Carlsson at Stanford.

I met Gunnar in a reading course I signed up for during my second year of grad school. These are directed studies where, typically, a faculty member guides a student through advanced/seminal research papers in their area. Gunnar is a widely respected mathematician with deep contributions in pure algebraic topology. During one of said meetings—we were reading Wall's *Finiteness Conditions for CW-Complexes*—I asked Gunnar about what he was actually working on those days. He proceeded to tell me all about this new field, topological data analysis, showing that it was possible to leverage machinery developed by pure algebraic topologists, but now in algorithms to solve real problems in data science. It made total sense, and I was blown away. We stopped reading Wall's paper, Gunnar became my thesis advisor, and from then on I've been working in TDA as my main research focus.[3]

---

[3] Here is a recent talk (in Spanish) from the *Cibercoloquio Latinoamericano de Matemáticas*: youtube.com/watch?v=cBJOo0NX6So.

I graduated in 2011, and we were able to arrange for my parents to make the trip from Cali to California. I believe that was the first time they had flown on an airplane. We had to get passports and visas, I flew back to Colombia and then came back with them for the commencement; having them there was awesome. After the ceremony in the math department, where we actually got our diplomas, Gunnar came up to me with congratulations and said he wanted to meet my parents. I panicked for a second and said: Gunnar … they don't … speak English—to which he responded: it's ok, I just want to say hi. He goes up to my parents and tells them, slowly, in Spanish, *José es un muy buen estudiante.*[4] He then said bye and walked away. At the beginning of this chapter, I mentioned the idea of small acts of kindness from mentors, that can have a huge impact. The interaction I just described is a prime example of one.

## Post-Stanford Years—Making the World a Smaller Place

After Stanford, I went to Duke University as a postdoc to work with John Harer. I was very excited to learn from and work with him—John has made pioneering contributions in algebraic topology, especially in the homology and cohomology of mapping class groups, and he has translated this expertise to establish foundational results in TDA. What gave me pause, however, was moving to North Carolina without any other context than the painful history of the overt racial violence that has plagued the South. As an outsider, the history books were my only reference. My experience there was positive, though. I always felt welcomed, and the four years I spent at Duke were fruitful both academically and personally. John was a great mentor to have; he helped me grow into the next stage of my professional career, and he and his family welcomed me into their home for every celebration.

After four years at Duke, I moved to Michigan State University (MSU) as an Assistant Professor with joint appointments in the Department of Computational Mathematics, Science & Engineering (CMSE), and the Department of Mathematics (MTH). At MSU I've worked to build a vibrant research group with undergrads, grad students, postdocs, and the support of several sources (e.g., MSU, Defense Advanced Research Projects Agency (DARPA), National Science Foundation (NSF) and its CAREER program), advancing applied topology in ways that I personally find deeply fascinating. Over the last few years, I've been fortunate to also mentor students (e.g., by co-directing their thesis work) from Mexico, Colombia, and Brazil. One of them is now completing a dual PhD in CMSE and MTH at MSU. With other Latinx faculty at U.S. institutions, we have also run summer schools, workshops, and programs both in the U.S. and South America, which we hope to leverage in order to reach students that may not be aware of the many existing opportunities.

The emergence of new areas in computational mathematics—like applied topology, applied algebraic geometry, computational harmonic analysis, etc.—provides a valuable opportunity to increase the representation of historically underserved communities in the mathematical sciences. Indeed, these areas blend mathematics with modern science in ways that students tend to find compelling, and with mostly democratized points of entry. Nowadays, academic software is open and widely available, online conferences and open

---

[4] José is a very good student.

courses can be followed anywhere in the world, and online social communities provide valuable information and mentoring that not too long ago were available to only a select few.[5] If you know of someone that you think may benefit from any of these resources, please reach out to them; a seemingly small act of kindness can go a long way.

## Advice

I would like to end with a few words of advice:

(1) *Advice to my younger self*: Trust in your own ideas and pursue them with unrelenting passion; even if it doesn't seem like it at times, things will work out in the end. Also, the tenure track is a gruelling process and busyness is a reality, but you must make time to take care of yourself both physically and mentally. Learning how to say "no" early on, and abandoning perfectionism, are good ways to create that space.

(2) *How to encourage and inspire younger people in math*: At least when I was younger, there were two things that motivated me to do mathematics. One was the deep personal satisfaction I would get from solving a difficult problem, and the other was hearing the stories and learning about the human side of being a mathematician (Simon Singh's *Fermat's Last Theorem*, is one of those great stories). With this in mind, one of the best ways I have found to inspire students in mathematics is to facilitate their own discovery, with ample guidance, but avoiding easy/lazy answers that would deprive them of the satisfaction of figuring something out on their own. Something that I also think is valuable is connecting the act of doing mathematics to the social aspect of being a mathematician. By this, I mean the stories, the struggles, the joy of sharing with people at conferences and other social/cultural events. Also, math is just cool, with intrinsic beauty and many connections to science broadly; try (as tempting as it may be) not to make it uncool with unnecessary jargon and over-formalization.

---

[5] See, for instance, my lectures on TDA at youtube.com/watch?v=APgR3avai30 and the *Network of Minorities in Mathematical Sciences* group on Facebook.

# 19

# Dr. Angel Ramón Pineda Fortín

## The Grandmothers (Abuelitas)

My personal mathematical story begins with two women who changed our family history. They supported their children to be educated in ways that no one in their family had been before. My maternal grandmother (*abuelita* Amada) was a single mom who raised two daughters (my aunt Gloria and my mother, Guadalupe). *Abuelita* Amada studied to be a secretary, but chose to support my mother through medical school by having a small store out of her house (a *pulperia*) in Tegucigalpa, the capital city of Honduras. The entire store was just a small room of her house facing the street, but *abuelita*'s hard work in that store changed the

Dr. Angel Pineda

Illustration created by Ana Valle.

direction for our family. When my parents were still in medical school, with two young children, we lived with *abuelita* Amada. Later, when my parents were established doctors and my two youngest sisters were born, *abuelita* Amada came to live with us. *Abuelita* Amada dedicated her life to her daughters and her grandchildren.

On my father's side, *abuelita* Alejandrina (Nina) had eleven children and lived in San Nicolas, Santa Barbara, a small town in a coffee-growing region of Honduras. My father

Photos courtesy of Angel Pineda.

(L) Abuelita Amada and my mother Guadalupe. (R) Abuelita Nina with 10 of her children. My father, Angel Sr., is standing and the farthest to the right.

(Angel Sr.) was the seventh of those children and the first to wear shoes before he was school-aged. He only got to wear shoes that young because he was sickly and *abuelita* Nina was worried about his health. My paternal grandfather (*abuelito* Pancho) raised cattle and supported *abuelita* Nina, but the drive for education came from her. *Abuelita* Nina's drive to have her children educated and the sense of family collaboration she instilled in her children created the opportunity for my father to go to medical school.

## The Parents: Guadalupe and Angel Sr.

I greatly admire both of my parents. They were the first in their families to go to college, get a post-graduate degree, and study abroad. My mom became a doctor in Honduras and specialized in anesthesiology in Mexico. My dad also became a doctor and surgeon in Honduras and sub-specialized in hand surgery in Mexico. We are largely the product of our environment, and my love of learning came from a home where books were everywhere. My parents both encouraged my intellectual curiosity, but also balanced it with a love of sports and travel. At first, this story sounds idyllic, but it is amazing that it came to be.

My parents married during their second year of medical school and had me in their third. Because Honduras has an open admissions policy to medical school, the classes are very competitive and the completion rate is very low. It took a strong drive for them to both have graduated coming from modest means and while having children (my sister Marcela was born two years after me). By the time my dad went away to school, six of his siblings had already gone to school for trade schools or studied to be teachers. When the first siblings left San Nicolas to study, they left by mule because there was no other way. Even though they didn't have much money, they rotated giving us a stipend each month.

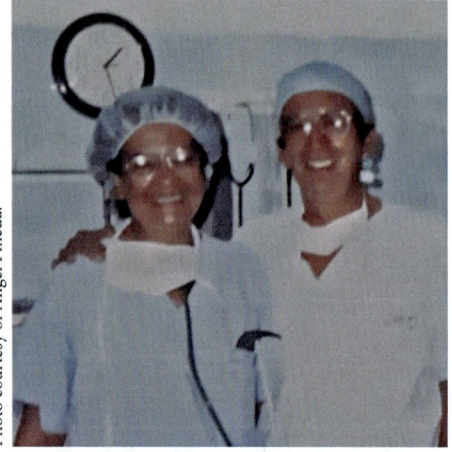

Photo courtesy of Angel Pineda.

My parents (Guadalupe and Angel Sr.).

This family collaboration continued with my dad's younger siblings who also went to college with the support of the older siblings. As a medical student, my dad worked providing healthcare to sex workers. My mom and dad would study with a child in one hand and a book in another. Both through hard work and family support, they were able to succeed. As I reflect on anybody's success, my parent's story highlights the importance of having a support structure along with hard work. As with everyone, my personal story began with the generations before me, who shaped me into who I am.

## Life Before Coming to the U.S.

I was born in Tegucigalpa, Honduras in 1972. From a very young age, I loved school, and mathematics in particular. As a young boy, I would "work" at my grandmother's *pulperia* at the register. Arithmetic was my first mathematical love.

When my parents were medical students, they didn't have much money, but they always prioritized school. We always went to the best school they could barely afford and sometimes we would be called to the principal's office because we were late in the payments. My parents always put my sister Marcela and I in bilingual schools so we would learn English, even though they themselves didn't speak the language.

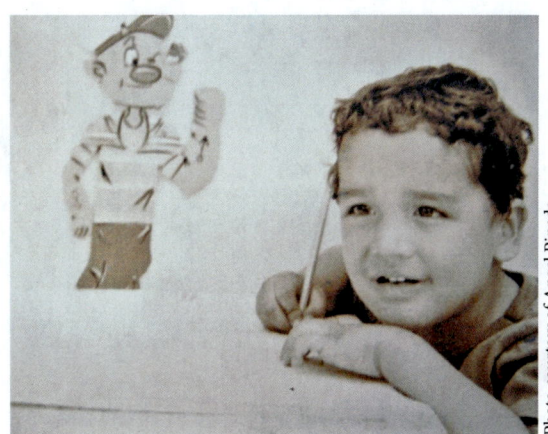

Kindergarten photo.

Photo courtesy of Angel Pineda.

We moved a lot when we were young, settling in San Pedro Sula, Honduras, when they began their medical careers. My parents and my sister Marcela still live there.

When the time came to apply for college, I didn't know anybody who had gotten as much financial aid as I needed to attend college in the U.S. Because of that, I started college in Guatemala since that was the best school we could afford. As a long shot, I also applied to colleges in the U.S. that I found had financial aid for foreign students by reading a book at the local library. Lafayette College accepted me and gave me more financial aid than both my parents earned in a year. That library book on financial aid ended up changing the trajectory of my life.

## Life as a Student in the U.S.

When I first came to the U.S., I had never been to Lafayette College. An uncle picked me up at JFK airport and put me on a bus to Easton, Pennsylvania. On that bus to Lafayette, I had a financial aid letter, but I wasn't sure it was real. My high school grades were good, but my SAT scores were average for Lafayette. It was hard for me to believe I received that much financial aid.

At Lafayette, I studied chemical engineering. Chemical engineering is interesting and useful, but I also chose it because if I went back to Honduras it would allow me to get a good job. My favorite college classes were in math. I loved the universality and structure of the ideas. Like so many immigrants, financial security played a big role in my decision of which major to choose.

My time at Lafayette was a time of professional and personal growth. From a professional side, I learned that in the U.S., I could make a good living as a mathematician. I also found that I wanted community service to be a core part of my life. Here is where one of my strongest mentors and advocates on both of those dimensions comes into my story. Rob Root was my undergraduate mathematics research advisor and has become a lifelong mentor and friend. At every stage of my career, Rob has been a voice of wisdom and support who has helped me get to where I am.

I went on to the University of Arizona to get a PhD in applied mathematics. Arizona has an interdisciplinary program which was a good fit for a chemical engineering student

Photos courtesy of Angel Pineda.

(L) Rob Root (my undergraduate advisor and I at the 2010 Joint Math Meetings. (R) Harry Barrett (my PhD advisor) and I in his house in Tucson in 1997.

who only had a math minor. Even then, I failed to pass my PhD qualifying exams the first time. If I didn't pass it the second time, I could not continue in the program. This brings me to what I believe is one of the biggest factors in success: persistence. I worked even harder for the second test and passed the exam. Regardless of the outcome, I just wanted to know I had given it my all.

After passing my qualifying exams, I was in a course in medical imaging where I saw the professor describe the human body as a mathematical function and imaging in terms of estimating that function from finite measurements. These thoughts combined my love of math with my desire to use mathematics in way that would help society like my parents did in their work in a public hospital in Honduras. That professor, Harry Barrett, eventually became my PhD advisor. I have heard it said that the most important decision you make in your life is choosing your life partner, but the second most important is choosing your PhD advisor. I agree and I feel fortunate to have picked Harry. Both as a mentor and a friend, he is an extraordinary person who shaped my approach to science through his combination of theory and practical application; combining the aesthetic and functional beauty of mathematics.

Like most people in a PhD program, I struggled at times, but a constant was Harry's belief that all of his students, with enough time and support, could do great work. This belief in his students is something I internalized and try to impart to my own students today. I particularly have this view for students who underestimate themselves. Unfortunately, underrepresented students are often overrepresented in having this mindset.

## Professional Career

After graduating with my PhD I went on to work as a postdoctoral fellow at the Radiology Department at Stanford University. It was a wonderful experience where I learned to apply the theory I knew from my PhD to clinical problems bringing me closer to patient care than I had ever been. In one of the projects, I was part of a team which included radiologists, physicists and engineers developing a method for separating water and fat in magnetic resonance imaging (MRI) in collaboration with GE Healthcare. This project resulted in four patents and is currently being used in MRI scanners. To me, it is amazing that a

Photos courtesy of Angel Pineda.

(L) Research students at California State University, Fullerton. (R) Research students at Manhattan College.

theoretical tool (Cramer-Rao Bound) could contribute to clinical care of patients. Being part of that effort was my biggest contribution to medical research to date.

Even though I was coming from a postdoc at Stanford and had several publications, it was difficult for me to get a tenure-track position. The first year I applied to mostly research institutions and I didn't receive a single interview. The second time, I applied to schools that had a balance of teaching and research that was closer to my own balance. That time I received three interviews and two offers, choosing to join California State University, Fullerton (CSUF). CSUF was a great fit for me. It is a Hispanic Serving Institution with many of the students being the first in their families to attend college. CSUF also has a master's program in applied mathematics which emphasizes industrial applications. This student demographic and mathematical breadth made it a place were I could pursue both my interests in social justice and medical imaging research.

At CSUF I mentored undergraduates in research, first through NIH funding and later through a grant from the Center for Undergraduate Research in Mathematics (CURM). That experience led me to become a member of the MAA subcommittee on research by undergraduates. I also became involved in mentoring underrepresented students through the Math Alliance. Continuing on the path of mentoring led me to be a joint principal investigator (PI) with Scott Annin of the NSF-funded Graduate Access to Research in Mathematics (GRAM) program which provided support for underrepresented students to attend graduate school. Like so many successes before it, it was not funded on the first attempt. Once again, persistence and collaboration were two major ingredients in making GRAM happen.

At CSUF, I also developed my teaching philosophy. As a new assistant professor without much teaching experience, I sat in the classes of teachers who were known to be effective. I learned that one has to teach in one's own voice. As a moderately introverted person, my voice was not as entertaining as some effective teachers. However, I found that I taught best in a quiet way, but using the curiosity about the material which also fuels my research.

After getting tenured and promoted to associate professor, I fell in love with a long-time friend, Tanya, whom I first met when we were research students at Lafayette. I had been single for a long time. A friend once said I was looking for a unicorn as a partner, a person who didn't exist, but I found her. Tanya lived in New Jersey and had two school-aged children who became my kids. Once again, I was looking for a job and this time on the other side of the country.

During my job search, I was fortunate to receive an offer from Manhattan College, a small Lasallian college in the Bronx. Manhattan College serves a large number of low income students and was starting a master's program in applied mathematics—data analytics. It was a great opportunity for combining social justice and applied mathematics. As the story repeated itself, the importance of fit in getting a job became clearer to me.

At Manhattan College, I keep working on teaching, research and service with the emphasis changing over the years. Currently, I am excited about the new NIH grant, of which I am the PI, to use neural networks to accelerate magnetic resonance imaging by assessing image quality by how well we can detect a signal. Once again I am able to combine mathematics, mentoring students, and improving medicine into one project. The hours are long, but the work feels like play. I am fortunate.

## Supporting Mathematics in Developing Countries

I often think of the people in Honduras and other parts of the world who do not have the same access to education as we do in the U.S. Ideally we would all have the same opportunities regardless of where we are born.

The first time I applied to the volunteer lecturer program (VLP) of the International Mathematical Union (IMU) I was not selected. Later, during a trip to Southeast Asia with friends, I visited a volunteer lecturer in Cambodia. The next year, I was selected to teach graduate numerical analysis at the Royal University of Phnom Penh (RUPP). As part of an international team, we were helping to rebuild the mathematical community that had been wiped out by the genocide brought about by the Khmer Rouge. That began a path of service associated with the IMU that so far has included co-authoring a report on the state of mathematics in Latin America, serving on the Commission for Developing Countries, and currently being the secretary for Graduate Research Assistantships in Developing Countries (GRAID), a program providing research assistantships to students in the poorest countries.

Photo courtesy of Angel Pineda.

Last day of classes after volunteer teaching in Cambodia.

## Being an Immigrant

I have a perspective that is rooted in my upbringing growing up as a person in the racial majority in Honduras. About 90% of Hondurans are *mestizos* (a mix of Native American and European ancestry). Here in the U.S., when I fill out the census form, I check both the Native American and White boxes because *mestizo* is not one of the options. Having grown up as a majority shaped the way I look at the world in a way that is different from many of my Latinx students who grew up in the U.S. I have the mentality of the majority even though I am a minority.

I am fortunate to have the language and culture that helps me connect with my Latinx students, but at the same time I realize that I have had many advantages they have not. When something bad happens to me, I rarely wonder if it is because I am brown even if that may be the reason. Growing up, my mother thought I could be president of Honduras, and I did too.

## Family

I love my job, but what I love the most is my family and friends. Aligning my time with my values is at the heart of my issues with work/life balance. I find that the best way to find balance is with those we love. My wife and I run marathons, and we say that life is a marathon, not a sprint. Even if you love your job, it is easy to burn out. It is important to find joy in all aspects of life and align your time with your priorities. This is a work in progress for me.

Photo courtesy of Angel Pineda.

With my three sisters (Marcela, Emma and Denise), *abuelita* and my parents.

Photo courtesy of Tanya Brubaker.

With my wife (Tanya) and two kids (Max and Alex) at a sea turtle rescue in Costa Rica.

## Advice

As I reflect on what I wish I had known earlier, I think of three things: we all need help, don't give up, and align your time with your values. Throughout my life there have been people who have helped my family and me. An example that resonates is when my dad's siblings helped support him in his education. When my younger sisters came to study to the U.S., I tried to do the same for them. If possible, it is important to surround yourself with a supportive community and to support others.

I also think of how many times I failed before I succeeded. There are so many examples of that, but failing to pass my qualifier exam at the PhD level really stands out for me. I could have been happy in a different career path, but I really love what I do. Persisting, not giving up, made the difference. It is important to try to learn from failure and keep trying.

Finally, I am currently working on being deliberate about aligning my life choices with my personal values. Within work, I value teaching, research, and service and try to align my time with how I value these aspects of my career. More broadly, on an almost daily basis, I try to create boundaries for my work so that I have a balance with my personal life. It is important to find a balance that feels right to you.

# 20

# Dr. Hortensia Soto

## In the Beginning

I want to begin with the story of Agustin Soto and Sara Ramírez, my parents, because their journey, their struggles, their sacrifice, and their work ethic molded me into the person I am today. It's important that I apologize to my parents because I am omitting their life experiences that occurred before they were married. They both came from well-to-do families; my mom was the tenth child of eleven children, and Daddy was the oldest of eight children. They each had their own adventures; my mom recalls climbing trees, going to *serenatas*,[1] while learning all the skills needed by Mexican women of that era, which included grinding corn for the *masa*,[2] cooking, sewing, raising children— all things to be a "good wife." Daddy's life consisted mostly of work; at age six he plowed with an ox, and at the age of eight he spent nights out in the fields, where the corn bundles scared him. He credited his *Padrino*[3] Jesús, who was very patient with him, for making the work bearable and fun.

Dr. Hortensia Soto

Illustration created by Ana Valle.

My daddy Agustin and mom Sara married on October 4, 1962, in Santa María del Valle, Jalisco. They lived next to my paternal grandparents where Daddy farmed and raised pigs with his dad. While they each had a wealth of life skills, they only had a third-grade education. Their marriage started off fairly normally until my paternal grandfather was shot and killed. After this tragedy, my paternal grandmother sold everything, including Daddy's animals, and moved to Nochistlán, Zacatecas and left my parents with nothing. This dramatically changed the direction of my parents' lives.

My parents then moved to Belén del Refujio, Jalisco where Daddy worked for a farmer. This is where my older sister, Eliasar, and I were born. Photos below show the two-room

---

[1] *Serenatas* are musical performance delivered in honor of someone or something.

[2] *Masa* is Spanish for dough that is traditionally made of flour used to make tamales, tortillas, and many other Mexican foods.

[3] *Padrino* is godfather in Spanish.

Photos courtesy of Hortensia Soto.

(L) Daddy, *Tía* Tila, *Padrino* Pablo. (R) Mother, my maternal great-grandfather Manuel, and *Tía* Eva.

home where I was born; the piece on the end is the *hogón,*[4] where my mom cooked. In this adobe home, my mom saw snakes coming through the roof as my sister and I slept. My parents suffered much during this time. They struggled to feed us, and Daddy almost died because he became ill and didn't have money for a doctor. He used to talk about his sandals that were held together with corn husks. Hunger led us to emigrate.

My maternal grandmother's dying wish was for us to move to the United States, where my mom had a lot of family members. My mom's brother, *Tío* Lupe, who lived in Nebraska, contracted the *coyote*[5] and we were set to cross the Rio Grande without either of my parents knowing how to swim. It was 1967, and I was a little over a year old and Eliasar was three years old. Daddy crossed first, with his total savings of $7.00.[6] My mom

Photos courtesy of Hortensia Soto.

The house where I was born—one door went into the bedroom and the second to the kitchen; I visited it for the first time when I was 15. I am on the far right, mom on far left, boy in blue is brother Mauro. Others are cousins and *Tía* Anjelina (Daddy's sister).

---

[4] *Hogón* is the Spanish word for bonfire.

[5] A *coyote* is a common term for a person who is hired to assist people to cross the U.S.-Mexico border.

[6] For reference, we note that based on inflation rates in the United States, $1.00 in 1974 is approximately the equivalent of $5.56 in 2020.

was supposed to be next, but fear set in and she refused to cross. This resulted in the three of us staying with the *coyote*'s grandmother for ten days at the border; there I learned to walk, while Daddy waited in Nebraska for the *coyote* to figure out how to get us to him without having to swim across a river. For $30.00, the *coyote* arranged for my mom to use a woman's passport and for Eliasar and I to use passports from another couple's daughters. Although there were passports for all three of us, crossing the border had to be done in shifts—first my mom and then me and my sister. The *coyote* took my mom across and left her at a theatre (this was her first time at a theatre!) and instructed her to stay there until someone came by to notify her that we would be arriving. With fear that she would never see us again, my mom waited about five hours. Finally, a young man arrived and escorted her to the back seat of a car, where my mom was uncertain of what would happen. Then in the rear-view mirror, she saw a car emerge, the *coyote* walked out, opened the door, and there were her girls. After driving straight through, only stopping once in Colorado for a meal, we reunited with Daddy. My mom and dad raised us to always give thanks to God for the people who have helped us; *Tío* Lupe and his family are on that list.

We settled in Morrill, Nebraska, where the population numbers fewer than 1,000. Daddy had odd jobs and one of his first big jobs was helping to build the Morrill Golf Course. His bosses were impressed with his attention to detail, his ability to learn quickly, and his innovativeness so much so that they recommended him to a local farmer, Art Dienes, who was seeking a hired man. Art wasn't keen about hiring Daddy because he didn't speak English, but my Dad said he would work for free for two weeks, and then Art could decide whether

Top: Mother and Daddy after reuniting in U.S.
Center: Eliasar and me after arriving to U.S.
Bottom: Agustin Jr. on the way.

Photos courtesy of Hortensia Soto.

or not to hire him. At the end of the first week, Art decided to hire him. This job allowed us to move from a house that shook when the trains went by to a house where Daddy made all of us sit in the car for fear that the house would explode when he lit the pilot to the heater. We later moved to another two-room house. I remember this house fondly because Eliasar and I played in a run-down school bus that sat behind the house. By this time, my brother Agustin Jr. was born, and we discovered that we had been reported to the immigration authorities. Art helped Daddy get a lawyer, who let us know that we couldn't be deported because Agustin Jr. was a U.S. citizen.

After a long wait and another addition to the family (my brother Ernie), we finally obtained our green cards and were allowed to live in the U.S. legally. By this time, I was in kindergarten. Also, when my mom was pregnant with Ernie, Art and his family moved to a smaller house so we could live in the main farm home. This was

Top: Me and my mom on my second birthday in the house that Daddy feared would explode. Bottom: Another farmer, Art, and Daddy.

Photos courtesy of Hortensia Soto.

a house with indoor plumbing, a restroom, a phone, a front and back porch, an upstairs, a dining room, and a big yard. According to my five-year-old eyes, we lived in a mansion. We moved out of that house when I was 12, but to this day when I dream of home, it is that house. Art helped Daddy start farming on his own, shared his equipment with Daddy, and was instrumental in Daddy buying our farm (this is a magical story). Art became our grandfather we called him "*el patrón*"—not in reference to being the "boss," but to the patron saint who saved us. He and his family are on the gratitude list.

## Hortensia's Early Years

I did not know English when I started kindergarten, but with the most compassionate teachers, I slowly learned. In honor of some of my first teachers, I want to share a few memories. I clearly remember learning the word "scissors" when my kindergarten teacher taught me how to cut because I didn't know how to hold scissors. My first-grade teacher left my name up on her door so I could go to it to see how to spell my name—it took me a while to learn this task. She didn't make a big deal about it; she let me do what I needed in order to learn. That year, Daddy also discovered that I had memorized my reading books and didn't actually know how to read. He quickly put a stop to that by randomly selecting words for me to pronounce, covering all the other words, and making me sound out words one at a time. During this time, he attended night school to learn English thus he was able to help.

I have two very vivid and important memories from second grade, one in spelling class and one in math class. After spelling "lace" correctly out loud, my teacher asked me to use it in a sentence. Knowing that I was at a total loss, she added "it's on your dress." My mom made all my clothes and there were lots of things on this dress, such as buttons, polka-dots, and a zipper. None of these resulted in a positive response as I pointed to them, so my teacher hinted that it was at the bottom of my dress. I incorrectly translated bottom to mean under, so I slightly lifted my dress to show my slip, but with excitement my teacher commented: "you are touching it." That day I learned the definition of lace and I also learned what patience looked like. The second memorable experience from second grade was the day we were exposed to exercises that looked like this:

$$3 + 5 = \square, \quad 3 + \square = 8, \quad \square + 5 = 8, \quad 11 - 7 = \square, \quad 11 - \square = 4, \quad \square - 7 = 4.$$

I struggled with exercises of the last two types, but I found them intriguing. I wondered how one would get the answer without trial and error, which is what I did. Imagine my excitement when I learned algebra—memories of these exercises flooded my brain and I was in awe.

Second grade might have been when my passion for mathematics began. Most people who know me know that my fifth-grade teacher transformed my life. The teacher kept me in during recess to catch me up so I could move to the "high group"—the group of students who were more successful academically. I was not excited about this because the "high group" did not have any Hispanics and according to me, since they could afford to be in band, they were also rich. I cried as my teacher walked me to the "high class," the class for the high achieving academic students; she hugged me and said that I would be just fine. I was worried about feeling out of place, but my first class with the "high group" was mathematics and she was right, I did just fine. By seventh grade, I decided that I wanted to attend college. Knowing that this was only feasible if I got scholarships, I decided that I would work towards becoming valedictorian of my class.

Even though they were different worlds, it is difficult to separate my educational experiences from my home experiences, so I will try to weave the two worlds. While most kids yelled with glee at the end of the school year, I cried because I hated summer. Yes, I hated summer! At the age of six, I started working in the fields, hoeing beets, weeding beans, and even weeding cornfields. I did this until I went away to college. My summers consisted of getting up at 4:30 AM, packing a breakfast, getting the younger kids ready, helping my mom prepare lunch, and going to the fields. We generally arrived between 5:30 and 6:00 AM and ate breakfast there; the kids who were less than six years old stayed by the car. My mom was creative and covered the windshield of the car with a blanket and left the doors ajar, so it would stay cool. We went home for lunch at noon, washed the dishes, and by 2:00 PM we were back in the fields till about 7:00 PM; sometimes we had to go irrigate the fields after this. It wasn't unheard of for us to have dinner at 9 PM. This was our routine six days a week, starting mid-May until school started in August.

In the fall, we had other harvest-related work; in the winter we mostly helped to separate calves from the cows or move cattle. I didn't complain much about the work in the fall and winter, but of all of us nine kids, I complained the most about the summer work. It seemed that every farmer north of Morrill wanted us to work their fields, and I wondered why their kids didn't do the work. My mom frequently reminded us (mostly me) to be grateful because these farmers were trying to help us; they knew that we needed work. My parents worked so hard, especially in the summer; I can only imagine what time they arose in the morning. The one benefit of working in fields was that it gave me time to daydream. I daydreamed of becoming valedictorian, going to college, becoming a lawyer, and helping Daddy pay off the farm. Unlike other kids, I didn't learn to swim, play sports, go to the movies, go to birthday parties or have friends spend the night. I learned to work.

Given that I didn't have a social life, it was easy to bury myself in learning during high school, and my mathematics class quickly became my favorite class. Math was the last homework that I worked on—it was dessert. I had the same mathematics teacher all through high school, and his pedagogical knowledge was ahead of his time. He rarely lectured; instead, we worked in groups on scaffolded packets where we discovered the big ideas. Sometimes, we had oral exams where the teacher probed further into our under-standing. I loved my "aha" moments, where I connected concepts and explained them to others. I got pretty good at explaining and if there was a need for a substitute teacher, I was asked to teach the mathematics classes for that day. This seemed crazy to me because it meant missing other classes, but I loved it. For me, mathematics was one big puzzle and each class offered more pieces to the puzzle. My teacher was very supportive and encour-aged me to pursue mathematics as a career, but I wanted to be a lawyer and most impor-tantly, I hadn't yet convinced my mom that it was OK for a young woman to leave home to attend college. This was a huge obstacle!

My mom's belief was that girls stayed home and learned how to become a wife, until they got married. As I got older, I understood this, but as an adolescent, I fought with her quite a bit about this issue. I was stubborn and determined to go away (far away) to college —I would not work in the fields for the rest of my life. Art had Daddy's ear and shared the importance of a college education. In fact, Art and a woman whose house we cleaned set up a scholarship for Eliasar to attend one of the local community colleges (another magi-cal story). This was my breakthrough; in my junior year, Daddy said that if I started off at a local community college and lived at home, then he would help pay so that I could finish my bachelor's degree. I agreed but knew that this would be a financial burden because by this time there were nine kids in my family. Thus, I continued working on my valedictori-an goal and decided that I would become a naturalized citizen when I turned 18 so that I could qualify for Pell Grants.[7]

I graduated high school as prom queen, president of student council, and valedictorian,

---

[7] Pell Grants are a subsidy the U.S. federal government which provides financial support for students who need it to pay for college.

Photo courtesy of Hortensia Soto.

Top L to R: Diana (deceased), Ernie, Agustin Jr., Bruno, Mauro. Bottom L to R: Eliasar, Daddy, Mother, Sarah, and me. I have another sister Norma who died when I was 12.

and that summer Eliasar and I became naturalized citizens. My first two years at Eastern Wyoming College (EWC) were completely paid for with scholarships. I was on my way to becoming a lawyer. As a side note, all my siblings have a college education. When my youngest sister, Sarah, graduated from college my mom looked at Daddy and said, "We did it; they all have an education."

## Hortensia's College Life

At EWC, I started off as a political science major and was the only mathematics tutor. After completing first-semester calculus, I met with my advisor to discuss courses for the following semester where I planned to enroll in Calculus II and the following conversation occurred.

**Him:** Calculus II isn't a requirement for a political science major.
**Me:** But we didn't finish the book.
**Him:** Don't you think you should be a math major?
**Me:** Yes, I do.

After graduating from EWC, I moved away to start summer classes at Chadron State College (CSC), where I planned to become a high school mathematics teacher because I had no idea what else one did with a mathematics degree. With new scholarships that covered tuition at CSC, I just needed rent, food and book money, so within a week I was a Pizza Hut waitress. At CSC, I also graded for the most amazing and supportive advisor and teacher, James Kaus, who is on the gratitude list. In his classes, he challenged us, we struggled and worked together, he patiently asked questions, and we learned. One day, while working on a topology problem in his office, he remarked, "You should get a PhD." I asked, "What's a PhD?" I don't remember doing anything special—I was just working away on the problem that I asked him about. I valued and trusted Mr. Kaus—if he said to do something then I did it. I didn't even know what a PhD was so I didn't have any goals on getting one. I only wanted one when he suggested I get one, when I realized he believed in me.

I student-taught, but I knew that wasn't what I wanted to do. So during a job fair at CSC, I visited Mr. Kaus and told him about my uncertainty of teaching high school. He suggested that I apply to CSC's master's in mathematics education program, so I did. I received a position as a Teaching Assistant that included tuition and a $3,000.00 stipend a year. This wasn't enough money to cover my living expenses, so I tutored on the side and I also worked at a local store on the weekends. Teaching collegiate-level mathematics immediately felt right—I found my passion. I loved my time at CSC, because I got to spread my wings. I am also very proud of my CSC education and that my parents didn't have to pay for any of it. If I needed fun money, Eliasar was my bank. Eliasar was working at an insurance agency at the time. We were raised with the philosophy that the more you give the more you get. Eliasar claimed that every time she lent me money, she would get a raise or bonus. After earning a bachelor's and master's in mathematics education, I was in debt $400.00 to her … interest-free. She is on my gratitude list.

Upon graduating from CSC, I applied for a job as the Director of the Mathematics Learning Center at the University of Southern Colorado (USC). One of the interview questions was about where I saw myself in five years. I replied, "working on my PhD." I got the job, and this is where I met my future husband, who was a statistician in the department. Shortly after getting married, we moved so I could pursue my PhD in mathematics education at the University of Arizona (UA).

**Struggle**. For some reason at UA, I immediately felt inferior. It seemed that all the other graduate students came from elite schools and that I was under-prepared for what lay ahead. Some of these students had already earned a PhD in another country. Every mathematical concept seemed so foreign to me. For the first time in my mathematical life, I was scared that I wasn't smart enough, and I lost all my confidence in my ability to learn mathematics. The fact that by the third week some of the courses dwindled down in size, scared me even more—if the smart people dropped the course, what was I doing there? Although the graduate students supported one another, I felt no support from the faculty, and they didn't seem eager to create a rapport with any of us. One time when I asked a question in analysis, the instructor replied, "All I can do is say it louder…" and then he said it louder. When I went to his office to ask about a homework problem that had been marked incorrectly, he said, "I don't really think you know what you are doing so, I didn't read it," as he flung my homework back to me. After 18 years of having some of the most compassionate, patient, and encouraging teachers, I was at a place where teachers quickly dismissed my questions. I did not pass one of the analysis qualifying exams and, thus I had to leave the program. My advisor helped me to find a new program in mathematics education and it happened to be at the University of Northern Colorado (UNC). I left UA with a second master's in mathematics, deflated, embarrassed, feeling less than human, and certain that I didn't know any mathematics. It took me years to get over this.

**Resurrection**. The first faculty member that I met at UNC was the graduate advisor, Dr. Ricardo Diaz (also on the gratitude list). On our first meeting, we were scheduled to go over courses that I had completed, and I was unconfident and full of shame. His first words were, "you have a very strong math background." These words brought a little glimmer of hope that maybe I could earn a PhD. I cruised through the program and surprised myself at what I knew. All of a sudden, I had ideas, answers, and creative proofs—my confidence came back. Yes, I struggled with some concepts, but I wasn't afraid to tinker and my instructors were helpful, pushed me, and had faith that I could do it. In retrospect, I did learn a lot of good mathematics at UA and developed as a mathematician, though it wasn't clear at the time. Most importantly, at UNC I learned how to conduct mathematics education research. I defended my dissertation shortly before my thirtieth birthday, received the Dean's Citation for Excellence Award at graduation, and delivered the commencement address. I did it!

## Hortensia's Professional Life

My professional life has been full of wonderful surprises. After completing my PhD, I started as an assistant professor of mathematics at USC and was accepted as a Mathematical Association of America (MAA) Project NExT fellow. This is a two-year professional development program for collegiate faculty that offers suggestions on integrating student-centered teaching and learning, writing grant proposals, and maintaining a research program. As part of this program, the second-year cohort creates and offers sessions to the new cohort. At the end of my first year, one of the project directors, Chris Stevens, asked if I would organize the sessions for the new cohort. I was stunned and honored that of the 70+ fellows, she asked me and, of course I said yes—which is what Joe Gallian taught us as part of Project NExT. This is how I found my professional home. In 2002, I became the first Project NExT fellow to serve on the MAA Board of Governors. I can still remember seeing Martha Siegel (Secretary), Anne Watkins (President), and Tina Straley (Executive Director) on the stage running the show. It was the first time I saw women with such power and authority—I was in awe. Martha and Tina quickly took me under their wings and invited me to serve on committees, some of which I was not qualified for, but they believed in me. The MAA community seemed to see something in me that I didn't know I had: leadership skills, which they nurtured and continue to nurture. I am beyond grateful for all the opportunities that the MAA has offered. It has been a pleasure to serve on the various committees, as the Governor for Minority Affairs, Associate

Teaching with an embodied activity.

Photo courtesy of Hortensia Soto.

Photo courtesy of Hortensia Soto.

Receiving the MAA Deborah and Franklin Tepper Haimo Award for Distinguished College or University Teaching of Mathematics.

Treasurer, and now as the Associate Secretary. In fact, I am the first female, first Hispanic, and first mathematics educator to serve in the role of Associate Secretary. I truly LOVE this community, which consists of so many friends.

After spending nine years at USC, now known as Colorado State University—Pueblo, my then-husband and I decided to leave USC because I wanted an opportunity to conduct more research and I had a high teaching load there; thus, I began to apply for jobs. UNC was also hiring, but I did not apply because I was certain that they wouldn't hire one of their graduates. Later they called me, asking me to apply, and I got the job. The downside was that I had to give up tenure and rank, but I did get three years of service towards promotion—this was very stressful. I spent the next two years working till 2 AM and sacrificed family time. Teaching 18 credits a year, conducting my own research, and guiding dissertations took a toll on my marriage. I got divorced (so grateful that we remain friends), became a single mom, and delved into work. My research on the teaching and learning of complex analysis along with my work on embodied cognition began to thrive. My graduate students are publishing in top-tier journals—professional life is good.

My work has been recognized by both UNC and the MAA. At UNC, I was the recipient of the College of Natural and Health Sciences' *Excellence in Faculty Research Mentor* at the Graduate Level and the *Excellence in Service Award*. I am also the recipient of the *Burton W. Jones Distinguished Teaching Award—MAA Rocky Mountain Section*, and the *MAA Meritorious Service Award*. I am also the first Hispanic person to receive the *MAA Deborah and Franklin Tepper Haimo Award for Distinguished College or University Teaching of Mathematics*. In fall of 2020, I started as a tenured full professor at Colorado State University. My work has paid off—I am so grateful.

## Conclusion and Advice

Although I didn't know it growing up, I am blessed to have a big family. My parents instilled in us a work ethic and a strong faith in God. From this story it is probably no

Photos courtesy of Hortensia Soto.

(L) Daddy and me on his 74th birthday. (R) Me and Miguel.

surprise that I am a daddy's girl—he passed away in 2017 and I miss him dearly. He was the first to hear of any of my successes. My son Miguel Agustin Johnson is my greatest gift. We are very close, and I treasure any time with him because he is pure love.

My advice to students is do not be afraid to have dreams that seem unreachable. The people who believe in you will emerge and push you to become more than you dreamed. My advice to mentors is that sometimes it is the little acts of kindness that make the biggest difference. Do not be afraid to be a human being and vulnerable with your students. Have high expectations while showing patience and compassion.

# 21

# Dr. Roberto Soto

## My Parents

The story always begins with my parents, Carlos y Maricruz. They immigrated from Guatemala in the 1970s and were looking for opportunity, both for themselves and for the children they hoped to have one day. And indeed, by 1990 they had five children, two boys and three girls, of whom I am the oldest. My parents placed a high value on education because they themselves desired it, but were limited by their circumstances. My dad is the eldest of ten siblings and had many responsibilities as a young man and only studied up to the eighth grade. My mom on the other hand worked since she was five years old and had to stop her education at the sixth grade.

Dr. Roberto Soto

Illustration created by Ana Valle.

At eight years old, my dad helped my grandfather by planting and harvesting beans and corn in the family plot. The harvest was meant to supplement the family's main source of income—my grandfather's tailor shop. My dad also started helping out in the shop at age twelve, was sent to the city to apprentice at thirteen, and by fourteen years old he was serving customers by himself. At nineteen years of age, my dad opened up his own shop and was trying to complete his education at the same time. Yet it was difficult to do so, thus two years later, in early 1973, my dad made the journey to the U.S. to set the foundation for our family's future, even though he was not yet married to my mom.

My mom, on the other hand, is one of seven and only two of her siblings had the opportunity to study beyond the sixth grade. Unfortunately, my mom was not one of them despite being good with numbers at a young age. By the age of seven, my mom would take a ten-mile bus ride to buy cheese at a cheese factory and then take another thirty-mile trip to the city to sell the cheese at the market. She also needed to make it back home by noon to eat lunch and attend school from 1–6 PM. Thus, at the age of seven, my mom knew how to mentally calculate change, manage credit, and navigate a complex system of transportation! By fourteen, my mom was working full-time at a textile shop in downtown Guatemala City, where she eventually met my dad. By 1974, my mom was ready to pursue bigger opportunities to continue helping her parents while joining my dad to get married and start a family of her own. To this day it is difficult for me to fathom that she traveled

Photos courtesy of Roberto Soto.

(L) A picture of my parents and me in 1976. (R) A picture of my family in 1998, our parents Carlos *y* Maricruz, and my siblings, Leslie, Angel, Angelica, and Rosa, from left to right.

through two countries by herself at seventeen! My mom is tenacious and a main reason for our success.

My parents ended up living near downtown Los Angeles. Fortunately, my dad's skills as a tailor and my mom's experience at the textile shop allowed them to quickly find full-time jobs in the textile industry of L.A. As they navigated the culture of Los Angeles, they also spent a lot of time worrying about *la migra*[1] and had a few anxious moments. But by 1975, my parents were able to become permanent residents because they had a child born in the U.S.—me. In turn, my parents became the gateway for the rest of their family to also enjoy the freedoms and opportunities that the U.S. offered. There are now many college graduates and business owners among my cousins and their children because of the risks my parents took to gain a better life for the entire family.

Compared to my parents, my siblings and I have lived a life of privilege. Not only did we have the basics in life, but we were able to pursue extracurricular interests and not worry about having to support our family as our parents supported theirs. Yet it was not easy for us to navigate our journey in the U.S. We still had the challenge of pursuing an education in a school system that underserved and continues to underserve our communities.

## Early Life

Although I remember that I liked going to school, I also remember that my parents couldn't always help me with my homework. This probably came from my earliest memory of struggle—I didn't understand what a certain first-grade assignment was asking me to do and my parents didn't know enough English at this point to help me. I remember feeling devastated that my parents couldn't help me and that I was not going to be able to complete my homework. I must also admit that the only parts of math that excited me in school were the timed exams on basic computational skills. This probably came from the fact that my parents enjoyed playing number games with us and encouraged us to be quick with mental math. In fact, the only two school math-related memories that

---

[1] *La migra* is slang for immigration enforcement in the United States.

I have are the following: On the first day of school, my fifth-grade teacher, Ms. Tanner, told us that she loved math and my sixth-grade teacher told us that if we divide by zero, the answer is zero. My sixth-grade teacher's mistake was caused by a question of mine. I remember that she took out a calculator, punched a few numbers, and told us the answer was zero. As you can imagine, this affected me greatly when I took calculus in high school and I didn't figure out what she did until I played with the district's basic calculators as a teacher fifteen years later. When you try to divide by zero, the calculator screen does show a zero, but it also shows a capital E in small print next to the zero to indicate that it is an error.

At home, mathematics was more exciting and deeper than what I learned at school. For example, as soon as I learned how to count, before I turned five, I started counting every car I saw on the freeway when we visited Tijuana, Mexico. When my mom asked me why I was counting the cars, I told her it was because I wanted to get to the last number in the universe. This pursuit of finding the "last number" continued once I was in elementary school. About the time I was in fourth grade, our parents bought us a book that contained math puzzles and stories. One particular story intrigued me. It was about a girl who was hired to rake her neighbor's leaves, and the neighbor offered to pay her in one of two ways. She could either receive $100 a day every day for a month or she could start with a penny today, two pennies tomorrow, four pennies the next day, and so forth doubling every single day for a month. The book then asked the reader to determine the best choice for the girl. I remember not looking at the answer because I wanted to explore this problem myself and I was surprised to discover that getting paid a penny on the first day and then doubling every day would be the best option. This "doubling" story also reminded me of the problem that I really wanted to solve which was to discover the "last number" of the universe. I decided on a new "strategy"—I would start at 1 and double each number until I found the "last number" I kept filling out paper after paper with numbers and my mom just let me compute because it was keeping me out of trouble. Finally, she asked what I was doing, and I told her that I wanted to get to the last number of the universe and that I had "discovered" a new technique that would allow me to get there faster. My mom looked at me quizzically and left me alone again as I filled out more sheets of paper with ever increasing numbers. Finally, she felt bad for me because she told me that there was no such number. I was shocked and I remember doubting her quite emphatically. She then challenged me to think of the largest number that I could and I said a googol times itself a googol times—I had recently learned from this book that a googol was a one with 100 zeros behind it, $10^{100}$. She told me that she could think of a number bigger than mine and I was dumbfounded! How could this be? She replied by stating that she could add one to the number I gave her, and my mind was blown! In fact, she insisted, she could always add one to any number I could think of and so this meant that there could be no "largest number." I fell in love with my mom's argument—my first proof by contradiction—and I should have known then that I was going to pursue math as a field of study.

Nonetheless, my family was not aware that we could pursue math as a field of study, and they did not possess the means or knowledge to encourage or provide me with the opportunities that could have helped deepen my passion for mathematics. Even a few more math books could have made a difference, but my parents trusted that school would provide for all of my educational needs, including the mentorship that I needed.

Unfortunately, there was nothing at school that intrigued me as much as this book or my mom's knowledge of mathematics, and this was part of the reason that later in life I wanted to teach in my community. I wanted our students to have their curiosity encouraged the way my mom encouraged mine.

Regardless, by sixth grade, I motivated myself by trying to show the world that Latinos could perform as well as anyone at school. It was going to be a proof by existence, not knowing that mathematicians like Bill Velez and Rodrigo Bañuelos were already opening doors for me. My dad kept encouraging me to show the world that Latinos were more than how they were portrayed in the media. Thus, my mission became to excel at everything, from math to English to baseball. I was going to outwork everyone and earn the best grades, even though there was nothing in my previous academic record to suggest that I could. But by the end of sixth grade, I had earned straight As, except for handwriting. I was so upset that I focused on handwriting all summer long so that I could earn straight As in seventh grade, not knowing that handwriting wasn't graded in junior high. But it didn't matter—in fact, it strengthened my resolve to be the best.

Math kept capturing my attention. For example, in eighth grade, I remember walking home one day with my friends, not listening to their conversation because I was so intrigued with the idea of the commutative property of multiplication and how it was so succinctly encapsulated in the symbols $ab = ba$. I remember mentally multiplying different numbers to confirm this statement and how awed I was by the power of symbols. I also loved to prove theorems in geometry—they felt like puzzles, and it reminded me of my book. However, what I needed most and wasn't receiving was guidance in developing my curiosity in mathematics.

But not all was lost—I was lucky to have two adults at Sierra Vista High School (SVHS) who really cared for my well-being and who provided the mentorship I needed to fulfill my dream of obtaining a higher education. One was my English teacher, Ms. Kelly, who intimidated us, but who really wanted the best for us. The other person at SVHS who took care of me was our career counselor, Ms. Dunn. Both Ms. Kelly and Ms. Dunn believed in me more than I believed in myself. In fact, Ms. Kelly and Ms. Dunn would team up on me so that I would fill out scholarship applications. In one particular instance, Ms. Dunn took me out of Ms. Kelly's class because I had not turned in a scholarship application that required letters of recommendation from community leaders. I disqualified myself because I personally didn't know any community leaders. Ms. Dunn was exasperated with me and called my mom to ask her if she could secure the letters I needed. Ms. Dunn then sat me down so I could finish the application. By 3 PM mom had secured the necessary letters and was at school to pick me up. Ms. Dunn ensured I finished my application and sent us on our way to the post office to turn in the application by the deadline, 5 PM. It turns out that because of my involvement in our choir at Saint John the Baptist Catholic Church and my parents' involvement in a family ministry, my mom was able to secure a letter from our pastor and from a city councilman. Thanks to Ms. Kelly, Ms. Dunn, and my mom this scholarship and many others guaranteed that I had less than $10,000 in debt after I graduated from UCLA.

Looking back, I wish I would have had this type of mentorship in math and science. Although I took all of the advanced science and mathematics courses that our school offered, no one really tried to encourage me to pursue either as a career. I wonder if we, sons and daughters of immigrants, weren't expected to pursue these types of careers. We were definitely not exposed to anyone who "looked like us" who pursued mathematics in college. Thus, even though I never owned a computer and did not know what computer science meant, I still chose computer science as my major when I applied to universities. In 1993, there was no way that I was seriously contemplating a math major.

## The College Years

As a computer science major at UCLA I had the opportunity to participate in a program for historically underrepresented groups—the Minority Engineering Program (MEP). If I were to credit an entity for my success in completing my bachelor's degree, it was definitely MEP.

One of the features of the program was our problem-solving sessions led by a graduate student in STEM. The graduate student provided us with problems that were more challenging than our homework, and we were encouraged to work in groups to solve these problems. We always performed well on exams, and, to this day, I try to incorporate elements of these problem-solving sessions in my classrooms. MEP also helped us secure summer internships as freshmen, and I received an opportunity to work at IBM in northern California. That summer was important for two reasons. I was finally able to experience life in the U.S. outside of the L.A. metropolitan area. Unfortunately, the experience also reinforced the way my computer science classes made me feel—very inadequate. Many of my classmates owned computers and had been programming for years while I couldn't afford a computer. It was discouraging to see my classmates finish projects ten times faster than me. I tried to continue as a computer science major, but I was enjoying my math classes more than my computer science classes. Instead of seeking out the mentorship I needed, I decided to switch from computer science to math.

I definitely do not recommend that anyone do this.

My first quarter as a math major started off on the wrong foot. I was taking linear algebra, and I did not form a study group nor did I seek help from my teaching assistant (TA) or professor—I decided that I could handle this class by myself. But, when my TA returned my first homework assignment, I was really contemplating switching to history. My TA wrote something akin to, "we do not write two-column proofs in upper division mathematics." I received a zero for that assignment, and I remember being upset when I received a 61/100 on my first exam. The curve meant that my D would count as a B, but I was still wondering if I had made the right choice.

In any case, at this point in my career, I obtained a job at the Academic Advancement Program (AAP) as a tutor for an Introduction to Linear Algebra course. AAP focused on providing support for underrepresented students in the College of Letters and Science, and the culture of AAP provided the support and community I needed to succeed. In particular, tutoring the Intro to Linear Algebra class forced me to learn how to read a math book. Not only did this help me become a better tutor, but I also learned how to prepare for my math classes. I applied this technique to my own linear algebra class and I was astounded to find out that I didn't need to take as many notes in class because I already understood about 50–70% of the material. The lecture answered all of my questions, and I ended up earning an A. In fact, I earned an A in every math class except Probability and

Statistics, and at this point even though the evidence suggested that I could pursue an advanced degree in math, I was still unsure.

## Career Path 1

I enjoyed tutoring—it felt good to help someone with mathematics. Since I wasn't aware of my career options, I decided to give teaching a shot. Our math department had an arrangement with the College of Education where you could start their credential and master's program as a senior and have your credential and master's one year after graduation. I asked my math advisor for her opinion and confirmed that it was a good plan. She also mentioned that my GPA would ensure that I would be accepted. I sometimes wish that she had said that my GPA also made me a good candidate for a PhD in mathematics.

So I set out to ask for a letter of recommendation from my differential equations professor, and I was stunned by her response. She said that it would be a "waste" if I became a high school teacher. I was upset because she was reinforcing my parents' feeling: they also thought that I would be wasting my talent as a high school teacher. But I truly felt that I could be an agent for change if I became a high school teacher in Baldwin Park to change lives by empowering students with mathematics. It took me a while to understand what my professor and parents meant, but I'm glad they shared their thoughts with me.

I have never forgotten what my differential equations professor said because deep down I knew that she wanted the best for me. In any case, one benefit of having earned a master's in education was the fact that my eyes were finally open to the injustices so many of my brothers and sisters faced around America and in my hometown of Baldwin Park. Those two years spent learning about education in America helped me understand that I was "lucky" to have gotten this far. I also learned that I could do more than just teach mathematics—I could use mathematics as a driver for social justice. Mathematics could open doors that would allow my brothers and sisters to fully participate in our society.

## Teaching is a Noble Profession

I loved teaching high school—I taught for a total of eight years—and I especially enjoyed the six years that I taught at my alma mater. I have enjoyed watching my students grow and become successful adults. They inspired me to be the best that I could be for them every single day so that they could pursue their dreams. They also challenged me to keep growing by asking me, "what comes after calculus" and what it meant to pursue a graduate degree in math. I didn't know how to answer, but I was fortunate to meet someone who was going to drastically change my life because she believed in me more than I did—Melissa, my wife of almost 14 years.

Melissa and I met in 2000, and she was so excited about math that she gave me a hug when she found out I was a math teacher. She also helped me reflect—I spent a lot of time helping students fulfill their potential, yet I had not done so. Moreover, she noted that I truly had a passion for math, was pretty good at it, and owed it to others to fulfill this potential. If I truly wanted my students to have access to the opportunities that were denied to our community, then I needed to explore how to gain such access myself and share my knowledge. It took me a while to gather the courage, but at thirty, I enrolled in a master's

Photos courtesy of Roberto Soto.

(L) Melissa and me singing a church song in 2001. (R) Getting married in 2006.

of arts in math program at California State University, San Bernardino (CSUSB). A year later Melissa and I were married and I finished my degree in 2008. I took a small detour to earn a little extra money for our families in 2008, but by 2010, we were in Iowa where I was set to start the PhD program in mathematics.

People always ask me how someone from Baldwin Park ends up in Iowa City. Frankly, I finally found the mentorship that I needed. Melissa introduced me to her advisor and LSAMP director at CSUSB, Dr. Belisario Ventura. I was searching for a PhD program that supported their students the way that MEP had supported me at UCLA. The first school he mentioned was the University of Iowa. Melissa and I arranged a road trip and in 2009 visited the University of Iowa where we met Dr. Phil Kutzko. Phil invited us to return to the Field of Dreams conference a week later and sure enough Iowa students were able to confirm Dr. Ventura's recommendation.

At 35, I enrolled at Iowa and started fulfilling a dream that I could never imagine. Having life experience made it easier to navigate the difficulties of a PhD, but the community at Iowa was also instrumental. Melissa and I met many friends who became part of my support network as we completed the journey together. Many professors at Iowa cared about our well-being, and I was blessed that my first algebra professor at Iowa was Dr. Frauke Bleher. Frauke's class confirmed that my favorite topic in math was abstract algebra, and I could not have asked for a better research advisor. Frauke helped me grow

Photos courtesy of Roberto Soto.

(L) Melissa and me at my graduation from the University of Iowa in 2015. (R) With my parents on graduation day.

as a mathematician, a writer, a presenter and was instrumental in making sure I conducted research on specific finite-dimensional $k$-algebras.

Dr. Kutzko also helped me navigate the guilt I felt for pursuing pure mathematics as opposed to applied mathematics. My parents had instilled in us a desire to serve our community, which was easy to do as a high school teacher. As a pure mathematician, I felt that the math I was producing was not truly helping my community. But he reminded me that first of all, the act of obtaining a PhD in pure mathematics meant that I was helping to hold the door open for other minorities to walk through. Moreover, we never question the contribution that art and music have on society, and, likewise, pure mathematics has something to offer humanity. But finally, becoming a professor of mathematics would also enable me to return to my community and share what I had learned on my journey.

## Career Path 2: CSUF and Beyond

I am now an assistant professor of mathematics and math education at a place that is a great fit for me—California State University, Fullerton. Most of our students are first-generation. We are both a Hispanic- and minority-serving institution. I meet many students who, like me, are navigating a university system that can sometimes feel daunting. Many of my students also love math, but are not sure what to do with their degree. They like the idea of graduate school, but are not sure how they're going to finance it, not knowing that tuition is covered, and you also receive a stipend to earn a PhD. My current research also focuses on helping faculty develop their pedagogical skills to create classroom environments that are rigorous, inclusive, diverse, and equitable. And, although I have published a few papers in mathematics, I see that my time as a mathematician is now better spent in helping young mathematicians find their path.

In the last five years, I have advised nine undergraduate research projects with nineteen students. My students and I have also collaborated in establishing a community of minority scholars that is focused on helping everyone rise. My students make me proud every day—for starting an inclusive research club called PRIME (Pursuing Research in Mathematical Endeavors) that focuses on opening doors for others, for overcoming many obstacles in their pursuit of excellence, and for teaching me that our feelings about math are as important as what we know in mathematics.

Our family is now growing—my wife and I just had a daughter—and I want her experience to be even more privileged than the one I had. But I also want her and her cousins

Photos courtesy of Roberto Soto.

(L) Our family reunion through Zoom during COVID-19 in 2020. (R) Melissa, Analissa, and me in 2020.

to live in a world that values their lives, and the culture that they bring to the table. And I also want my daughter to learn not just from my experiences, but from those that my parents have shared with us. Our success has only been possible because of them and their bravery, tenacity, and work ethic.

## Advice

As I reflect upon my journey, if I could give my younger self advice, I would say "do not be afraid to ask for help." Although I learned to be an independent learner by trying to figure out everything by myself, it also left a blind spot—there are times when you need community to help you reach higher. It took me a while to learn this lesson, but I hope my story encourages others to seek a community of people who want the best for them and challenge them to fulfill their potential.

This is exactly who I try to be for my students. I invite them to be part of our mathematics community at CSUF. My first assignment for students is that they turn in a syllabus quiz in my office so I can get to know them and start breaking down barriers. I listen to their goals and aspirations, and I mention opportunities that might be of interest to them. I invite them to our PRIME Club meetings and I encourage all to join a club. And I challenge my students to overcome the doubts that creep into all of us. Many of my students wonder if they've made the right decision. I try to tell them that most of their choices are not right or wrong, but have consequences that we can learn from to become better.

This is how I try to overcome the challenges that come from being a Latinx/Hispanic mathematician and educator. The biggest challenge for me has been in reconciling my identities. I went through life thinking that I had to be two different people—a person at home who is different than the one at work. Indeed, the toughest challenge has been balancing my life as a son, brother, husband, and father with my roles as a teacher, mentor, and researcher. I overcome these challenges by trying to reflect as often as possible on the choices I have to make—reflecting on if I am at peace with the consequences of my choices.

# 22

# Dr. Richard A. Tapia[1]

## My Story–Made in America

Frankly, if you don't know me, you may be wondering why you should care about my life or my work. In a nutshell, I have succeeded—against all odds—well beyond what anyone, including me, but excluding my wife and mother, ever would have dreamed. In 2017, the *Houston Chronicle* featured what they called the 36 most fascinating individuals in Houston. There I was, proudly next to Simone Biles[2] and her four Olympic gold medals. I hold the highest academic position at Rice University—University Professor—only the sixth person to hold this position in the 100-year history of the university and of course, the first Hispanic. The Blackwell-Tapia Mathematics Conference and the Tapia Celebration of Diversity in Computing Conference are named in my honor. I have received eight honorary doctorates from prestigious universities and given eight commencement addresses.

Dr. Richard A. Tapia

Illustration created by Ana Valle.

I was elected to the National Academy of Engineering. President Clinton presented me with the inaugural Presidential Award for Mentoring in 1996, and in 2011, President Obama honored me with the National Medal of Science, the highest award given by the U.S. government to an American scientist or engineer; in 2014, I won the prestigious Vannevar Bush Award from the National Science Board; in February of 2017, the American Association for the Advancement of Science (AAAS) awarded me the Public Engagement in Science Award.

For all of these awards, I was the first U.S.-born Hispanic recipient. I say all this not to brag, but to convince you of my credibility. I marvel at what an individual can do in America. Yes, I have lived the American Dream, but it was without the apple pie. I call my

---

[1] Much of the material in this chapter had its origins in Dr. Tapia's forthcoming book *Losing the Precious Few: How America Fails to Educate Minorities in Science and Engineering,* Arte Publico Press University of Houston.

[2] Simone Biles is an American artistic gymnast and is the most decorated American gymnast.

Photo courtesy of Richard Tapia.

Receiving the National Medal of Science from the 44th American President Barack Obama (2011).

life story "Made in America" because even though I have a Mexican heritage, I was molded in America and feel American.

I embrace the word Tejano, popularized by the late Selena,[3] and closely identify with the Tejanos. After all, I have lived in Texas for half a century; however, I was born and raised in California, so I can never be accepted as a true Tejano. Hence, I am both an honorary Mexican and an honorary Tejano. I am proud to say that here I am in my early 80s and I consider the four papers that I have recently written in my late 70s the best papers of my entire career. That is the way we minorities are. We often start late (I got my PhD at age 30) and get better with age. I have always loved mathematics, and in return mathematics has been very good to me. It has given me wonderful opportunities and much satisfaction.

## My Family

My parents Maria Magdalena Angulo (Magda) and Amado Bernal Tapia came as impoverished children from Mexico to Los Angeles seeking an education. Times were hard, as they had to support themselves and were not able to obtain the education that they sought. My mother came to Los Angeles alone, and lived with and was influenced by a Jewish family from the age of 12 to the age of 19, when she married. A similar story applies to my father, but in his case the family was Japanese. My mother went no further than middle school, my father finished high school. However, my parents' educational dreams were fulfilled through their children. I have a twin brother, Bobby; a sister four years younger, Ana; a sister, Rebecca (Becky), who is seven years younger; and a brother, Steve, who is seventeen years younger.[4] Out of five children, four of us have undergraduate degrees and three of us have graduate degrees. I am a product of these Mexican parents, the city of Los Angeles, and the time period of the 1960s. My father worked extremely long hours—often leaving the house for work before we got up and returning after we had gone to bed. They were hard working, good people who came here for better lives and they found them. They certainly gave as much to this country as they received from it.

[3] Selena Quintanilla, regarded as the Queen of Tejano music, was a Mexican-American singer and songwriter.

[4] Bobby passed away in April 2020. Ana passed away in 2009.

Photos courtesy of Richard Tapia.

My mother Magda.                                    My father Amado.

Although my parents were so very proud of being Mexican—while their hearts probably remained in Mexico—they adapted well to the new world and definitely made it work
for their children. My father was inclusive, and everyone loved him. To this day my wife
Jean claims that he is one of the people she loved most. While not everyone loved my
mother as much as they loved my father, everyone respected her. She was well focused
and strongly directed. As I reflect on my mother's teachings, I summarize them as: 1) Be
proud; 2) Believe that you can (*sí se puede*); 3) Demonstrate good work habits; 4) Strive for
global excellence.

## School Days

In the winter of 1943, Bobby and I started kindergarten at Dayton Heights Elementary in
the Los Angeles Unified School District (LAUSD) and not far from our house in central
Los Angeles. We were well groomed, but our dress style was outdated and we looked different from our male classmates. Our teacher Marjorie seemed to be quite fond of Bobby
and me, and we felt it. The experience was a pleasant one, as we mostly listened to stories
and played games with the other kids. On the other hand, Mrs. Anson's first-grade class
was a different story. Bobby and I would not join in group singing or participate in any
oral activities. We were extremely shy and much more comfortable speaking Spanish than
speaking English, but the language of the class was 100% English. Mrs. Anson called our
mother to ask if Bobby, the quieter of the two of us, had any speech problems. Of course,
she said no and decided that we would start speaking English at home.

We moved to Torrance from central Los Angeles in the summer of 1946. While our
house had a mailing address of Torrance, California, it was technically in the Los Angeles
Strip, a narrow region that runs from Los Angeles to San Pedro. At the time, the Strip was
not well developed and was heavily populated with so-called "Okies."[5] We started school

---

[5] Okies refers to farm families displaced from Oklahoma and nearby states to California in the
1930s by the Great Depression and the Dust Bowl.

in Torrance (actually in Carson) in September of 1946 at Carson Street Elementary. The next year we were zoned to Halldale Avenue Elementary School in Torrance. As I reflect, I remember good things about Halldale. Our fourth-grade teacher was Mrs. Bentwood. She was the best teacher in the world. Realizing that Bobby and I were shy, she tried hard to bring us out. We were the only Mexican-Americans in the class (school). Whenever in reading or social studies we would encounter a Spanish word, she would ask either me or Bobby to pronounce it correctly in front of the class. When we would do math, she would excuse me from the new material saying that I already knew it and she would have me tutor the students who were behind. She made us feel good about who we were, perhaps for the first time in school.

Bobby and I later went to Narbonne High School in Lomita, California for both middle school and high school. Due to crowded conditions, the middle school went in the afternoon and the high school in the morning. So, Bobby and I were free in the afternoons during our high school days. We spent all our afternoons and evenings working on cars, reading hot rod magazines, and listening to early 1950s music. Our passion was cars and drag racing. Bobby became a world-class driver and was elected to the National Hot Rod Association (NHRA) driver Hall of Fame.

Narbonne was an extremely low-performing school, arguably the lowest in the district. In spite of my low-performing school, I did well academically, especially in mathematics, and was considered a star. Yet no one—no teacher, nor any counselor—ever suggested that I could or should attend college. Few students from Narbonne went to college. Low expectations have hurt so many potentially excellent minority students.

In eleventh grade, the American Mathematical Society promoted mathematics appreciation among local Los Angeles high schools by administering a mathematics test to each school. The schools with high participation rates would be acknowledged. During an assembly, Principal Barnett encouraged participation by saying that the individual with the best score at Narbonne High School would be acknowledged and given an award in an assembly in front of the entire school. This was exciting. I took first place. Mr. Barnett

Me and my '57 Chevy.

Photo courtesy of Richard Tapia.

called me to his office and said, "Congratulations, here is your pin." I immediately asked, "But what about the assembly?" He replied, "There will be no assembly." This deeply hurt me. I was so much looking forward to shining in front of the entire school. To this day, I am confused as to why there was no assembly. I even entertained the ridiculous thought that it was because I was a rare Mexican-American in an "Okie" school.

## Pursuing Higher Education

Since no one encouraged me to pursue higher education, after graduating from Narbonne in January of 1956 I went to work in a muffler factory. I was happy working in the hot sun next to an individual from Mississippi who told me "Richard, do not make my mistake. You are smart, go to college." By the end of summer I could take no more and ran off to Harbor Junior College (HJC) in Wilmington, California in September of 1956.

I have fond memories of Professor Friedman's calculus class at HJC. Several of my friends and I would sit in the last row. Every Friday we had a 20-point quiz. I would get 20 points, my neighbor would get 19 points, and the next neighbor would get 18 points, on down the line. Friedman explained the difference in scores by saying that the person that originated the answer got the most points, then the person who got it next would get second highest, et cetera.

During my time in junior college I met Jean, my future wife, on April 21, Easter Sunday, of 1957. She was 15, about to turn 16, and I had just turned 19. We dated extensively through the summer of '57 and in September, her mother sent her to New York to continue her ballet studies at New York City Ballet. In Christmas of 1957, a friend of mine and I, at the spur of the moment, decided to drive to New York to visit Jean. I got back late and missed an exam in Professor Friedman's calculus class. Since I had never received a B in a math class, he said that he would give me an A in the course if I scored 100% on his final. There were five questions, I easily did four. But I did not know how to do the fifth. So, I thought and thought and finally came up with an approach. Friedman was not confident that the approach was correct and gave me no credit for that problem. I asked him to tell me what was wrong with my approach. After a day or two, he gave me full credit saying I had rediscovered an old theorem that was not well known.

Professor Friedman was perhaps the best math professor I have ever had; he told me to not go to a state school, but go to UCLA after junior college; so I did, in the Fall of 1958. At UCLA they told me that I could have been accepted with scholarship help if I had applied out of high school. No one ever told me that or even hinted at that. But there I was at UCLA, what a wonderful occurrence. I was a math star in high school and in junior college, but not at UCLA. I was just good enough. I survived by working hard for the first time in my life.

My path to graduate school was not certain or smooth. As an undergraduate math student at UCLA, I was an A-B student, but mostly B's. I had no illusions of going to graduate school on the strength of this record. In my senior year, two of my classmates told me that they were applying to the UCLA Mathematics Department for graduate school. I knew that I had done better and had more mathematical talent than both of them, so I applied and was accepted.

But there was a problem. I had married as a sophomore and had a daughter as a junior, and Jean and I supported ourselves by working part-time. We were really broke. Moreover,

I had taken out several student loans for my undergraduate education, forcing me to delay my acceptance and work for a year and a half. In that time, I worked for Todd Shipyards in San Pedro, California, on a grant from the United States Navy's Bureau of Ships. The project consisted in using mathematics to define the surface of a ship. When the project finished in 1963, I returned to UCLA for graduate school. The Todd Shipyard experience reinforced my view that I needed a PhD in order to take a leading role in interesting projects. The time I was in graduate school at UCLA, from 1963 to 1968, was an exciting period in U.S. history, especially in California. The 1960s was my favorite decade and it was as wild and exciting as people say it was. Until this time, I knew that I was Mexican-American, but at UCLA in graduate school, I became Chicano.

Of the nearly 300 graduate students in the UCLA Mathematics Department, I was the only domestic Latino. That did not really bother me because few Latinos attended my high school. When I started graduate school, there was one Black student. Naturally, we became friends, but he was a victim of the qualifying exams and had to leave UCLA. Classwork seemed fairly routine. I was not a star in class, but I was good enough.

For two years of graduate school, I had no financial support from the department, so Jean and I worked part-time, staggering our hours so one of us was always with our young daughter, Circee. I worked part-time as a supplemental (student) employee for IBM in Westwood Village near UCLA. And on weekends, I worked for my father at La Fleur nursery in South Gate, California as a salesperson. Jean worked as a PBX operator at Saint John's hospital in Santa Monica and taught social dancing at Arthur Murray Dance Studios. In spite of all this work, we had a great social life and attended parties weekly. Often, we would stay up all night partying and then go directly to work the next morning after going to a coffee shop for breakfast. I would take naps on the fertilizer stacks at La Fleur nursery.

My twin brother Bobby joined me at IBM. Our boss at IBM was a very charismatic and smooth senior executive named Joe Mount. Joe had a PhD in math from UCLA. Putting great importance on the choice of advisor—and rightly so—he directed me towards choosing his former advisor, Professor C.B. Tompkins, as my advisor. Tompkins had been an excellent mathematician in his day; however, he gave me no guidance on choosing, researching, or writing a thesis. I was left alone to choose a problem, do all the research, and write the thesis. Tompkins did not even know what my problem was about. However, Dave Sánchez, a member of my doctoral committee, entered into the later stages of the process and guided me in rewriting my thesis, "A Generalization of Newton's Method with an Application to the Euler-Lagrange Equation." At one point, I told Professor Tompkins that I was making little research progress on my thesis because I was working too many hours outside of the university. So, he went to Professor Magnus Hestenes, the director of the Office of Naval Research-sponsored UCLA Institute for Numerical Analysis and was able to obtain a research assistantship for me. I then quickly finished my thesis, a respectable contribution to the math literature.

## Struggles with Identity

As I grew up, I was often told by ignorant people "Mexican, go back to where you came from." When I visited Mexico at the age of 13, I was told "Gringo, go home." So my confusion led me to ask where do I belong, what is my identity? I am not a white American and I am not Mexican. At UCLA, in the mid 1960s, I would discover that my proper identity was being Chicano. That has served me well for my entire life. *Soy Chicano!* So the roles that I identify with and make me happy are: Chicano, Tejano, mathematician, and car enthusiast.

Put those four identities into one and you get something that looks like this image taken from the cover of SACNAS News.

When I was at UCLA there were foreign Latino graduate students and faculty, but we did not really understand each other. They did not understand me as

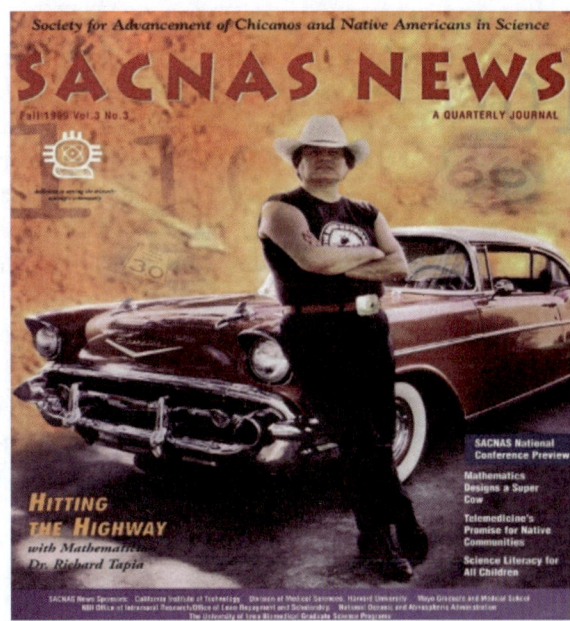

Cover of SACNAS News.
Photo courtesy of SACNAS (Society for Advancement of Chicanos/Native Americans in Science) sacnas.org.

a Chicano growing up in the United States and the resultant extra baggage. Until I met Professor David Sánchez, a 1960's Chicano, I believed that to be a successful math graduate student or math faculty member you had to be from another country. Sánchez was an appropriate and excellent role model for me. He knew well the path that I had traveled. We quickly bonded, and he served as the mentor and role model that I needed, but did not know that I needed. Foreign Latino faculty could not serve that function for me.

The Spanish language, which plays the dominant role in defining us Latinos, can also play a role in dividing us. When I first meet a foreign Latino, they invariably speak to me in Spanish. I can see that this clearly strengthens our newly formed relationship. However, my experience has been that this bond will sooner or later be challenged by my lack of proficiency with the Spanish language. Yes, I have been intimidated by native Spanish speakers all my life; hence, how can they serve as role models for me? I say this, not to air personal grievances, but to explain the complicated Latino identity issues that may impact mentoring among Latinos. As Latinos, we are not all the same and here is why it matters: If you hire a foreign Latino as faculty and think that he or she will be able a priori to mentor well your domestic Latino, that may not be the case at all.

Recently, one of our Chicano graduate students, who was born and raised in Los Angeles, decided to attend a meeting for Latin American graduate students at Rice. He is not fluent in Spanish, but at the meeting they all spoke Spanish and made fun of him, calling him a "fake Mexican." He said he made a big mistake and will never attend those meetings again. I completely understood him and shared his discomfort. To be an active member of the Chicano movement in the late 1960s you did not need to speak perfect Spanish. Maybe I take pride that my identity is formed, in part, by my deficiency in Spanish. Being Chicano asserts my identity. Being Latino does not carry the same impact because it is far too broad and weak a distinction. I have been scarred over this issue of fluency in Spanish. This makes me react in a defensive manner as if I have a chip on my shoulder, which I probably do.

Now, it is perhaps interesting to compare this impact that the Spanish language has on me with the impact that the language of mathematics has on me. If I walk into a room full of world-class mathematicians, I do not experience that same feeling of deficiency even though many of them will be far more proficient in the language of mathematics. But here I can stay quiet until we reach my area of expertise, where I can run with the world's best and may even lead the race.

Minority students are more likely to be inspired by those with whom they identify. Some believe that I am unnecessarily picky, but can't they see the importance of this? The experiences of URM[6] students place them in contact with non-URMs all the time. Although it would be unrealistic to assert that only URM faculty can be of value as mentors and guides for URM students, it is important to cultivate the mentorship pool to include mentors made from the same fabric as the URM students.

## The Professoriate and More

David Sánchez was the only domestic Latino mathematics faculty member at UCLA when I was a graduate student and he was on my doctoral committee. Upon graduation, he asked me what I was going to do. I replied that I did not know and would probably take an industrial job. He said that I should try academia. I had never thought of that, but told him that it sounded exciting, so I would consider it. Sánchez and Lowell J. Paige, the chairman of the UCLA Mathematics Department, called Barkley Rosser, the Director of the Army Mathematics Research Center at the University of Wisconsin Madison and convinced him to offer me a postdoctoral position. When I received the offer, I told Jean, by now my wife of nine years, to pack up our belongings and our two children because we were going to Wisconsin.

Spending the next two years at The Mathematics Research Center (MRC) was the best decision I have ever made in terms of my professional career. I was no longer a student and had an opportunity to run with the big dogs: world-class mathematicians like Barkley Rosser, I.J. Schoenberg, and Michael Golomb. They treated me like a colleague; I was no longer a student and running with the best in the world. When I went on the job market after being at the MRC for two years, I had multiple excellent offers from Tier 1 Research Universities. Jean and I accepted a position at Rice University in Houston, Texas. I have to wonder how many of us URM mathematicians would rise to the top of our fields if we are given opportunities like I was given at the MRC.

I started at Rice in 1970 as an assistant professor. At Rice I just wanted to be a good professor in terms of research, teaching, and service to the department and the university. However, I soon saw that as an URM who had traveled that challenging road, I could be very effective in mentoring and working with both undergraduate and graduate URM Science, Technology, Engineering, and Math (STEM) students. Moreover, there was a great need and there was no one who could step up to the plate. I immediately faced a critical decision that would challenge and plague me throughout my entire career. This is the

---

[6] Acronym for "underrepresented minority."

delicate balance between professional activities, mainly research, that would be rewarded and outreach activity that would not be rewarded, but was so much needed. We young URM faculty had heard many alarming stories about minorities not being promoted for various reasons.

It became clear to me that I should get tenure before I start doing significant outreach, so I did. Nothing tests an individual's survival skills better than figuring out the path to promotion with tenure and then following it. Too often, young faculty do not seem to possess or exercise the needed skills. I was really quite good at university survival. I received a promotion to associate professor with tenure in 1972, essentially in record time.

I always understood that you had to write many papers to secure tenure (they need not be great, but must be good enough to be published in good journals). You can write books and conduct high-risk, challenging, and truly important research after you gain tenure. As I reflect back, I probably advanced too early, but I made it, and it moved me even further towards the front of the bus. At times, I felt that I was driving the bus. Our chair suggested that I had such good visibility in the Rice community because I was an URM, and I had such good teaching evaluations because I had long hair. He may be right on the former point, but not on the latter point. I was now in a secure position to embrace giving back in terms of addressing underrepresentation. I can help because I have been there and I understand, and now I have tenure. The first thing that I did in terms of outreach was found the group Rice Association of Mexican-American Students (RAMAS) in 1972. In the photo below, you see our original group. We did have one woman in RAMAS, but she was absent for the picture. That was the gender balance in those days.

## Molding of Leadership–Rice Days

The maids, the janitors, and the groundskeepers were almost exclusively Mexican at Rice. They were so proud to see one of their own at the faculty level and they showed it in their respect for me, the first Mexican-American faculty member. They could not speak English, so they were overjoyed that there was a faculty member with whom they could identify with, and that they could talk to in Spanish. In contrast to the foreign Latinos that I described earlier, my bond with these women was outstanding; they were my people. One day several maids came to my office saying that there was something important

Rice Mexican-American Student Group.

Photo courtesy of Richard Tapia.

that they wanted to tell me—their boss, the director of buildings and grounds who was well-connected, was stealing from Rice. I asked these maids if they could have their immediate supervisor come and talk to me about this accusation. Mr. Cruz came and repeated the identical story. While I was trying to figure out what to do next, I was told by the maids that Mr. Cruz had been fired by the director of buildings and grounds. It seemed that he learned that Cruz had talked to me. At this time colleagues had shared with me that the German Department was trying to get rid of an excellent non-tenured young woman faculty member so that they could retain a not-so-excellent non-tenured male faculty member.

It was clear now that I must go to then Rice President Norman Hackerman with my two concerns. Hackerman was forceful, direct, and talking to him seemed like standing in front of an approaching Mack truck. Yet, he was a brilliant and well-recognized chemist and at the time he was the chair of the National Science Board. I decided to visit President Hackerman and relate my two concerns. My stories did not fall well with President Hackerman. He sternly told me that my stories could not be true and that I was an enemy of the university. I so clearly remember those words. I now realize this was a bold move, since I did not have tenure at the time, and furthermore, a few months before, I had a verbal confrontation with then Rice Provost Frank Vandiver at a general faculty meeting over some comments that he made in reference to minorities as faculty. About a week later, I was called back to the president's office. He told me that the German Department issue had been taken care of, that Cruz would be reinstated with back pay, that the top boss had been fired, and that he was going to nominate me to the National Science Board. He did nominate me, but I was not elected because I was junior faculty, and members of the National Science Board are distinguished scientists and administrators. However, I was elected some 25 years later. I realized that this was Hackerman's way of saying he respected my bold style.

Throughout my career I have been an active and visible leader at Rice. I was an active member of the undergraduate admission committee for six years. I was the chair of the Mathematical Sciences Department for five years, and I founded the President's Lecture Series of Diverse Scholars. I directed the National Science Foundation sponsored Alliance for Graduate Education and the Professorate for more than ten years. I founded the Tapia Center for Excellence and Equity in Education. In addition, I was a good citizen, good teacher, and good researcher. I gave Rice excellent national visibility in many components.

## Molding of Leadership—Beyond Rice

In 1968, the Chicano movement in Los Angeles and at UCLA was alive and strong. It was then and there that I found my identity. In 1972, New Mexico Medical School Professor Alonzo Atencio, with funding from the National Institutes of Health, called together a group of 17 young fresh science professionals to discuss the formation of an organization that eventually would be called Society for the Advancement of Chicanos and Native Americans in Science (SACNAS). We were brown with some shades of red, all male because that was the way that science representation was at that time. We desperately needed the support of each other, for only we, certainly not our university colleagues, understood the challenge of dealing with the extra baggage that we, as underrepresented minorities growing up in this country, faced in our professional life. Chicano gave me an identity, and SACNAS gave me a family that supported that identity. Our first meeting, in 1973, consisted of 50 young professionals getting together in Atlantic City, New Jersey. Over the years our membership grew and at the 2019 SACNAS annual meeting more than 4,000, mostly brown undergraduate students, attended. The early SACNAS members became more than my professional family—they became my family.

As an applied mathematics professional, I also became actively involved early in my career in the applied mathematics organization Society for Industrial and Applied Mathematics (SIAM). I attended its annual meetings, joined committees, gave talks, and served as conference organizer for several highly visible SIAM annual conferences and special conferences. My SACNAS and SIAM activities brought me nominations to prestigious committees, including the National Science Board, a Clinton presidential appointment. This visibility coupled with my research activity and well-recognized mentoring and direction of women and underrepresented minority students in turn led to prestigious awards, including selection to the National Academy of Engineering (first Latino ever) in 1992 and the National Medal of Science (only Latino ever) awarded by President Obama in 2011. This followed the creation of the David Blackwell–Richard Tapia Mathematics Conference in 2000 and the Richard Tapia Celebration of Diversity in Computing Conference in 2002. By now I had become a well-recognized national STEM leader with far more than my share of prestigious awards.

## Advice to Students and Mentors

It is said that what does not break you strengthens you. I surmise that I am not completely broken, but the personal tragedies of my dancer wife Jean battling multiple sclerosis for more than forty years, our daughter Circee's accidental death, and our son Richard's bouts with personal issues have caused me to live increasingly close to this boundary of being broken. I would trade my numerous awards and honors, and my wife Jean would suffer through multiple sclerosis again, to avoid the tragedy of losing our daughter Circee to an automobile accident. But we do not have that choice. Our only choice is to give up or play the hand that we were dealt. The choice is easy. Life has its strange twists.

When you encounter obstacles and adversity, learn to look both ways. Your challenge is to handle adversity. Prosperity is quite easy to handle. Realize that tragedy and failure are as much a part of life as are triumph and success. Failure is a part of every successful person's life. You must learn to grow from your failures and to develop compassion and sensitivity from your tragedies. At each stage of your life and career, continue to dream and work to make your dreams come true. However, learn to cope and still enjoy life if they do not all come true.[7]

---

[7] Part of this section is adapted from Dr. Tapia's commencement address at Harvey Mudd College in May of 2017.

# 23

# Dr. Tatiana Toro

## Early Life

**The family**. My father's family is from Antioquia, a department in northwest Colombia, lying mostly within the Andes mountains and extending toward the Caribbean Sea. The *paisas*[1] in my family are *mestizos*[2] mostly of Spaniard descent. My father, Gabriel Toro, is one of ten siblings. He grew up on a farm, where from the age of six he had to work alongside his brothers and his father as *jornaleros*[3] tending to the sugar cane crop. His mother and sisters cooked for the day laborers. My father attended school at most six months of the year when sugar cane was not in season. The first time he wore shoes was the day of his first communion; it was the same pair of shoes his brothers before him had worn for their first communion. He got his

Dr. Tatiana Toro

Illustration created by Ana Valle.

own first pair of shoes at the age of 17, when he left the farm and moved to Bogotá to start medical school at the *Universidad Nacional de Colombia*. Despite his struggles with food and housing insecurities, my father graduated from medical school at the top of his class. He obtained a fellowship to study neuropathology at Charles University in Prague. He crossed the Atlantic by sea during the summer of 1959. He returned in 1962 after obtaining a PhD and turning down a job offer in Prague and one in Havana, Cuba.

My mother's family is from Huila, a department in southern Colombia, spanned by the Andes mountains. The *opitas*[4] in my family are *mestizos* mostly of indigenous descent from the Yalcón *pueblo*. My mother, Gladys Calderón, is one of four siblings. She grew up mostly in Zipaquirá, a small town near Bogotá where her father was a school teacher. My grandfather Carlos Julio Calderón was the eldest of fourteen children. When his father

---

[1] A *paisa* is someone from a region in the northwest of Colombia, including the part of the Andes in Colombia.

[2] A *mestizo* is a person of mixed ancestry.

[3] *Jornaleros* are the equivalent of day laborers.

[4] An *opita* is someone from the Huila region of Colombia.

died of "sadness" during the great depression, he buried his wishes of becoming a physician and went to work to support his mother and his siblings. Gabriel García Marquéz, the winner of the Nobel Prize in Literature in 1982 was my grandfather's student in Zipaquirá. Gabo's dedication to my grandfather in his first novella *La Hojarasca* read: *A mi profesor Carlos Julio Calderón Hermida, a quien se le metió en la cabeza esa vaina de que yo escribiera.*[5]

My grandmother Carmen de Calderón was a school teacher who believed in education as a means to achieve success and who fought for the rights of her daughters to attend college. My parents, both first generation to attend college, met in medical school. They got together when my mother was beginning her residency in pathology and my father had just returned from Prague. I was born two years later under difficult circumstances. Because my father had been labeled a communist and despite being the only neuropathologist in the country, he was unable to find a job; he had been banned from most hospitals. My mother was very ill toward the end of the pregnancy and had a long road to recovery after my birth. My first few years of life were financially challenging for the family. By the time my brother was born, when I was almost four, the situation was a bit more stable.

**Grade school**. At age four, I started school in the *Lycée Français Louis Pasteur* in Bogotá. I was very lucky to have been accepted to the only private co-educational, non-religious school my parents could afford. The French government subsidizes these schools in developing countries to ensure that their citizens abroad have access to an education that is comparable to the one they would get in France. Having the opportunity to attend this school opened many doors and played a fundamental role in my deciding to study abroad.

When I think back to my grade school years I think of my best friend who I met in kindergarten 50 years ago. Our friendship started from the shame we experienced to have to wear suspenders. The school uniform required grey wool skirts. These were expensive so to ensure that they lasted a while our mothers had bought a bigger size and put suspenders to hold them in place. This brings me to the second thing I remember about the school—a deep sense of not belonging. The majority of the kids at school were from a very different socioeconomic class, the children and grandchildren of several Colombian presidents attended the school while I was a student there. The third thing that comes to mind are the math classes.

In first grade we learned set theory. We drew Venn diagrams on the playground and used the teacher's giant blocks (a huge magnification of our own set) to study unions and intersections. We were given a lot of freedom and I could not imagine a better math class. That year we also learned how to count in different bases. With a partner, we used wood structures that resembled buildings, small plastic boxes and beans to count in different bases and to translate between two different bases. It was wonderful.

We learned algebra in sixth grade from a beautiful book that I still remember fondly. At that time the French school system tracked students after ninth grade. I chose the math track. We learned calculus, linear algebra and some basic analysis in tenth, eleventh and twelfth (*seconde, première,* and *terminale*).

---

[5] The quote translates as "To my professor Carlos Julio Calderón Hermida, who had in his head the idea that I ought to write."

In 1981, the United States hosted the International Math Olympiad (IMO) for the first time. The U.S. decided to open the competition and invited several countries that had never attended before. Colombia was allowed to bring an eight-person team. The Colombian organizing committee invited a large number of schools in Bogotá to send up to four representatives to participate in the process put in place to form the team. By chance, I ran into the four boys (all in twelfth grade) that had been chosen to represent my school, as they were leaving to go to the first meeting. I asked the school if I could go with them. They told me they were only invited to send four kids, since I was in eleventh grade they did not include me. They told me that I could try to go as an individual, but that the school would not sponsor me. I went to the first session, explained my situation and was allowed to stay, participate in the training and in the qualifying examinations. I made the team. At that stage the school decided to sponsor me!

Although I did terrible in the competition which took place in Washington D.C., participating in the 22nd IMO in 1981 changed my life. I met a number of students who planned to study mathematics after high school; I did not know that was a life option. The French and the French-Canadian kids explained the path they were planning to follow, they were going to go to *les écoles preparatoires aux grandes écoles* (the preparatory schools for schools like *Polytechnique, École Normale Supérieure*, etc.) for two years, then take the exams for these schools and go study more mathematics. During those ten days I gathered as much information as possible: what the best preparatory schools were, what the application process looked like, what grades were needed on the *baccalauréat*, where people lived, etc.

Going to one of those preparatory schools became my main goal. For a Colombian girl in 1981 this was a science fiction scenario, not even a dream. I worked very hard to get the grades I needed in the *baccalauréat*, filled out the application and found out from the start that these programs heavily favored boys. For example, the schools provided boys with housing in the dormitories and forced girls to find housing elsewhere. Initially, I did not even pay attention to this detail, I was determined to make this work. I was accepted to a couple of places and decided to attend *Lycée Louis Le Grand* in Paris. Interestingly, the main roadblock I encountered was societal: in Colombia in 1981, young women were supposed to live at home until they got married. It was unheard of for a teenager to move to Paris on her own. My father's colleagues and siblings told him that no respectable father would ever allow such a thing to happen; it was considered the road to perdition.

## Higher Education

**France**. I turned 18 in the summer of 1982, so as an adult I did not need my father's official permission to leave the country. With my mother's support and against my father's wishes, I left for Paris. I was very naive, in my great scheme I had not considered the difficulties of living in a foreign country without any support system, especially coming out of a tight-knit family. Being able to only talk to my parents and my brother for a maximum of ten minutes twice a week was heartbreaking, first the anticipation and then the disappointment that the communication was bad and the time was too short. International communications were inefficient and extremely expensive, twenty minutes a week was all we could afford. My year in France was extremely difficult; I was very home sick. To my surprise, the school environment was dominated by deeply rooted male chauvinism. My thought was that after seeing the *machismo*[6] in Colombian society I had seen it all. I was wrong. In September, my class had 53 students, eight of whom were women. By the end of the academic year, only five women were left.

Some of the things I recall from the classroom conditions were that tests were returned in descending order of performance and the grades were read out loud. Students were regularly called to the board, loud disparaging comments were common as was the impassive attitude from the math teacher. I recall confronting him, in front of all the class for allowing this abusive environment. The physics teacher was a woman and I recall thinking at the time that it must have been very difficult for her to get where she was. My academic performance, which was deeply correlated to my emotional state, oscillated between very good and terrible. My support system included my aunt who lived in Sicily and who I was able to visit twice, and a dear friend, who I met at the dorm, and her family who basically adopted me. Thanks to their incredible support I managed to finish the academic year. I went back to Colombia for the summer wondering whether I would be able to come back for a second year. The experience had been grueling, my parents financial situation was dire and given my performance during the first year, the school placed me on the track to enter an engineering school. I was not interested in becoming an engineer.

**Colombia**. I returned to Colombia, took the entrance exam for medicine to the *Universidad Nacional de Colombia* in Bogotá and passed. As I was supposed to register to enter medical school I realized this was a mistake. It was not possible to transfer to the mathematics program that semester, but physics was an option. So I started my first semester as a physics student in the fall of 1983. This was very fortunate as I was able to immediately enroll in the physics lab that was required for math students in later semesters. In Spring 1984, I transferred to the mathematics program. On May 16, 1984, the *Universidad Nacional de Colombia* witnessed the bloodiest day in its history. After the kidnapping, torture and murder of a student from the campus in Cali, student assemblies and a demonstration were programmed for that day. Very soon the demonstration turned violent and the army was authorized on campus. As a result, students died and went missing, soldiers and policemen were injured, and the university was taken over by the military, surrounded by barbed wire and closed for over a year.

Those were dark times, I had a sense of having wasted a golden opportunity in France. I was unable to attend college as the National University (a public institution) was closed and we could not afford a private school. During that period I learned how to knit as a way to cope with stress, and I also started to learn English. My mother thought it might come in handy. I did not agree with her. Given the role the U.S. had played in Latin America through the years, I could not imagine why anybody would want to live in the U.S. I only agreed to start because my mother was making such an effort to keep me engaged and preventing me from sinking further into depression. I also decided to learn math on my own.

In the fall of 1984, faculty were allowed back into their offices. I wrote a petition asking that I be allowed to take exams for the courses in the major. I proposed to study on my

---

[6] A strong or aggressive masculine pride.

own (if possible consulting with a faculty member from time to time) and then, when ready, take an exam on the subject. A passing grade in the exam will amount to passing the course. My initial petition was denied, but with time they accepted. The only prerequisite to be able to do this was to have taken the physics lab course I had completed during my first semester. It was a lucky coincidence.

This is how I did most of my undergraduate studies. By the time the university reopened in 1985, I had passed a large number of the required courses to graduate with a degree in mathematics. I took math courses in person for a total of three semesters: one as a physics student, half before the closing and one and a half after the reopening. I had exhausted the offerings by then. A graduation requirement was to have been registered for four semesters. Fortunately, I was missing a history course, which was the only class I took my last semester. I received a BS in math from the *Universidad Nacional de Colombia* in December 1986.

Three of my professors encouraged me to pursue a PhD in the U.S. I knew I wanted to learn more math, although I had no idea what one did with a PhD in math. I also knew I did not quite fit in Colombia, and that I might do better somewhere else. The English lessons came in handy after all. Furthermore, it was more the hardships and the challenges that I had faced than the mathematics I had learned that prepared me to succeed in graduate school.

## Becoming an Immigrant

**Stanford**. I started graduate school at Stanford University in Fall 1987. I had assumed that there would be very few women in the program. Nevertheless, I recall being surprised during the welcoming event for the graduate students: there were 17 new graduate students and only one woman. Although less than 10% of the graduate students were women, the environment was healthier and more respectful than it had been in France. The graduate students were a very cohesive and supportive group. During the first quarter, in the complex analysis course, I met my future advisor, Leon Simon and my future husband, Dan Pollack. Leon was teaching the course, Dan was the TA. My English level required that I take a course to improve my spoken fluency. Dan used to joke that the homework for the English course was to find a *gringo*[7] boyfriend. I used to watch the news everyday as a way to improve my English. The first few months were exhausting.

My advisor was demanding, firm, and straightforward. He deeply valued hard work. These were all characteristics that suited my learning style and spoke to my work ethic. We all have moments of doubt along the way. In the summer of my third year in graduate school, things were not going well. I was unable to focus on research and therefore was getting nowhere with work. In the back of my mind was a nagging question: *What was somebody like me doing in a PhD program?* My country was falling apart and I was doing nothing to contribute to the community I came from. My life at Stanford felt artificial and meaningless.

---

[7] In Spanish-speaking countries and contexts, the word *gringo* is used to describe a person, especially an American, who is not Hispanic or Latino.

One day, toward the end of that summer, Leon asked me to go for a walk. He wanted to understand what was going on. He listened as I explained what I was thinking. He told me that from what he heard about the situation, he could infer that in Colombia I would be an easy target and that he doubted I would be listened to or given a chance to develop any of the ideas I had in mind. Then he told me that if I really wanted to help my country and my community I should become the best mathematician I could be, that that would give me the platform I needed. I am immensely thankful to him for that walk, for what he said and for how he said it. I have wondered many times what would have happened if we had not gone for that walk. Maybe this is a good place to mention that walking plays a huge role in my life. I walk an average of five miles a day. Some of the most important decisions in my life have been made while walking and some of the most important conversations I have ever had have occurred during long walks. I have walked sixty miles in three days for a good cause.

I obtained my PhD in 1992, after the fall of the Berlin wall in November of 1991 and the Tiananmen square protests in June 1989. These two events changed the world and the landscape of the job market for mathematicians in the U.S. I was fortunate to secure jobs that allowed me to continue through the academic path. Many of my classmates were not so lucky.

Through my experiences in Palo Alto and Menlo Park (two very affluent communities neighboring Stanford), as well as through interactions with some of the undergraduate students I taught, I learned that the color of my skin and the way I looked were considered appropriate topics of conversation, as well as reasons to assume I was uneducated and could be taken advantage of. This was a revelation, and something I have reflected upon through the years. I was 23 the first time somebody made it clear that I did not belong in her neighborhood and mocked me under the assumption I did not speak English. I have spent time trying to understand what happens to a person when this abuse starts as a child. I acknowledge that the reality of a Latinx individual born or raised in the U.S. can be very different from mine.

**Professional career.** After Stanford, I spent a year at the Institute for Advanced Study in Princeton, a year at UC Berkeley and two years at the University of Chicago. I met Carlos Kenig in Chicago. He has been a friend, a mentor and a wonderful collaborator through the years. In 1996 we moved to Seattle. I am half of a two-body problem[8] and the University of Washington (UW) offered us two tenure-track positions in a beautiful place, at a time when positions were very scarce. It was an offer we could not refuse. I went through the ranks at UW. I was promoted to Associate Professor in 1998 and to Full Professor in 2002. Initially my professional focus was in research. I followed Leon's advice to become the best mathematician I could be.

Over the past twenty-five years my research has developed in several distinct, but interconnected directions of analysis: partial differential equations (PDEs), harmonic analysis and geometric measure theory. The most representative theme of my research in PDEs corresponds to the theory that weak notions of regularity are well adapted to the study of boundary behavior of solutions to elliptic PDEs and to free boundary regularity

---

[8] The two-body problem in academia describes the difficulties an academic couple has in securing jobs at the same university or within reasonable distance from each other.

problems. The success of this program, initiated in Chicago with Carlos Kenig, has significantly expanded our knowledge in this field. It has solidified the theory that weak notions of regularity are suitable to study this type of problems, which thus far had only been considered in terms of classical notions of regularity. These ideas have opened a new area in analysis. They were a central theme of the harmonic analysis program at Mathematical Sciences Research Institute (MSRI) in the spring of 2017 and of the research term at *Instituto de Ciencias Matemáticas* in Spain in the spring of 2018.

I have been recognized with a number of prestigious invitations; I have been a speaker in the analysis session at the International Congress of Mathematicians (ICM 2010) in Hyderabad, India and in the 23rd Nevanlinna Colloquium, ETH, Zurich in 2017. I have given a number of named lectures, among others the NAM Clayton-Woodard Lecture at the Joint Mathematics Meeting in 2016 and the inaugural AMS Mirzakhani lecture at the Joint Mathematics Meeting in 2020. I am a Member of the American Academy of Arts and Sciences (2020), a *Miembro Correspondiente de la Academia Colombiana de Ciencias Exactas, Físicas y Naturales* (2017) and a Fellow of the AMS (2016). I have been awarded the Blackwell-Tapia Prize (2020) and the Landolt Distinguished Graduate Mentor Award from the University of Washington (2019). Furthermore my research has been continuously supported by the National Science Foundation since 1994.

I have had the good fortune to work with wonderful groups of junior mathematicians: graduate students, postdoctoral fellows and beginning assistant professors. I have made an effort to create vertically integrated groups where team members are both mentors and mentees. Seeing young people grow mathematically and flourish professionally has been very rewarding. They have brought me lots of joy.

I serve the mathematical community in different roles: as co-chair of the Scientific Advisory Committee at Mathematical Sciences Research Institute, as a member of the

Photo courtesy of Mariana Smit Vega Garcia.

Some of my mentees, who refer to themselves as "Toro-ites."

Board of Trustees of the Institute for Pure and Applied Mathematics (IPAM) at UCLA, as a member of the Board of Directors of the Banff International Research Station in Banff, and was a member of the Board of Directors of the Pacific Institute for the Mathematical Sciences (PIMS) at the University of British Columbia, Vancouver, Canada until early 2020. Currently I am a member of the U.S. National Committees for the International Mathematical Union. I was an elected member of American Mathematical Society (AMS) Editorial Boards Committee (2016–2019) and I currently serve as an elected member of the AMS Nominating Committee.

In winter of 2012, there were several CAMP students in the calculus class I taught. The College Assistant Migrant Program (CAMP) at the University of Washington is federally funded through the U.S. Department of Education's Office of Migrant Education. It is designed to reach out to and support students from migrant and seasonal farm-worker families during their first year in college. CAMP students are asked to get progress reports from the faculty after each test. The students started coming to my office because they were required to and then kept on coming. One of them talked to me after the final exam about how much it had meant to them to have me, somebody who looked like them, in a room with over 230 students with whom they did not identify. That same student told me that although she was not quite sure what a mentor was or did, she wanted to ask me to be her mentor. I am very happy to share that she is now a third-year medical student at UW, fulfilling her life's dream.

In 2013, in recognition of the small number of Latinxs in the mathematical sciences and motivated by my experiences with the CAMP students, I brought the idea of a conference for Latinxs in the mathematical sciences to Russ Caflisch, at the time the director of IPAM and to Alejandro Adem, at the time the director of PIMS. Their support and the hard work of my co-organizers brought my idea to fruition in 2015. The first Latinxs in the Mathematical Sciences Conference (LATMATH) was held in April 2015. Over 150 people participated, including undergraduate students, graduate students, postdoctoral scholars and faculty, and researchers in industry and government. Additionally, a group of high school students participated in a math circle and attended a panel featuring UW's President Ana Mari Cauce, among others. The CAMP student who inspired me to organize the event was also there. The participants expressed enthusiasm for another LATMATH Conference. The second edition took place at IPAM in March 2018 with over 250 participants. Sponsors included National Science Foundation, National Security Agency, Elsevier, the Mathematical Sciences Institutes through the Diversity Initiative, UCLA, Facebook and UW. I am extremely pleased that this conference is now one of the programs in the Diversity Initiative of the Mathematical Sciences Institutes. Latinxs are significantly underrepresented in math and science. There are many problems that contribute to this deficit, and the solution must be multifaceted. This conference is one step in the right direction. The third edition of the conference is scheduled to take place at IPAM in March 2022.

Issues of equity and underrepresentation are at the forefront of my professional interests. I believe that a solid scientific platform allows me to address these issues in settings where they are seldom discussed. To be successful we need to be represented all the way from the bottom to the top of the professional and academic ladders.

**Family**. This *testimonio*[9] starts with the family and ends with the family. As a Latina there is nothing more important than my family. I am very thankful to my grandparents, my parents, my aunts, my husband Dan  and my children Samuel and Sara for their love and support. Dan, Samuel and Sara have kept me honest, have inspired and challenged me. They have always believed in me. They have made my life richer and have given me the strength to always go forward.

---

[9] testimony

# 24

# Dr. Anthony Várilly-Alvarado

## Family History

Brazil has a special place in my heart: I owe my existence to a chance encounter in 1976 between a Costa Rican woman in her late twenties pursuing a master's degree in education, determined to make a better future for herself, and an Irish math PhD student who was following his advisor on their sabbatical at the *Universidade Estadual de Campinas*. After marrying in late 1978, my parents moved to Moravia, a suburb of San José, Costa Rica, where my mother had lived all her life until she went to study in Brazil. I grew up in a house built on the same modest plot of land owned by my great-grandfather; my father, Joseph, still lives there today. My mother lost a six-year battle with cancer in 2002.

Dr. Anthony Várilly-Alvarado

Illustration created by Ana Valle.

Her name was Jesusita, although most friends and family knew her as Susy. Jesusita de los Ángeles Alvarado Blanco, to be more precise. It is an unusual name, even by Latin American standards. My maternal grandmother, Julia Blanco Rojas, had been told she would have difficulty conceiving; she prayed and promised God that if she had a child she would name him Jesús. When my mother was born, Julia pivoted. My grandfather, Augusto Alvarado Montero, the oldest of five children, had many jobs in his life, including stints in his late teens picking bananas for the United Fruit Company on the Atlantic coast. He made good money doing this[1] much of which he sent home to help pay for the education of his siblings. He never got to go to college.

I didn't meet either one of my maternal grandparents. Julia died of ovarian cancer in 1965, when my mom was 15; Augusto followed in 1973, from a stroke. Julia had a second child, my uncle Enrique, but he died young, and so by age 24 my mother's immediate family was gone.[2] She had already completed a bachelor's degree in science education at the

---

[1] This is not to say that the working conditions were good; see the chapter *A la sombra del banano* in Carlos Luis Fallas's novel *Mamita Yunai*, originally published in 1940. (*Yunai* is a transliteration of *Uni*(-ted), the way in which most Costa Ricans referred to the United Fruit Company.)

[2] My mother had a stepsister, my aunt Nelly, who was 15 years older. For reasons that were never clear to me, they were not close in the 1970s. Whatever the rift was, time helped heal the wound, so I grew up with an aunt and older cousins whom I've always been fond of.

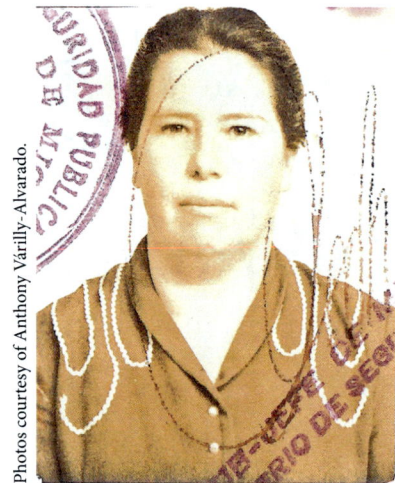

Photos courtesy of Anthony Várilly-Alvarado.

My maternal grandparents. (L) *Abuelita* Julia. (R) *Abuelito* Augusto.

*Universidad de Costa Rica*[3] and taught high-school chemistry at the *Liceo Vicente Lachner* in Cartago. A government scholarship allowed her to go to Brazil and get a master's degree; she jumped at the opportunity, and began learning Portuguese in preparation for the trip.

Picking up Portuguese as a Spanish speaker is not too hard. My father Joseph, however, did not speak Spanish in 1976. But he is a quick study, and he loves to learn new languages. His path to Campinas was in some ways even more unlikely than my mother's. He grew up in Dungloe, a small town in northwestern Ireland, the son of a policeman (and my namesake) and a homemaker, Nan Varilly (*née* Boyle). He is one of five siblings, all of whom worked hard to move up the socio-economic ladder from rather humble beginnings. By age 12, my father was out of the house, attending a boarding school in Letterkenny on a scholarship. Unable to attend Trinity College in Dublin[4] on account of being Catholic,[5] he studied mathematics at University College Dublin. There, he found a home and was well-supported by mentors like Seán Dineen. At age 21, he left his native Ireland for the United States, where he enrolled as a PhD student in mathematics at the University of Rochester. On Dineen's advice, he began working with Leopoldo Nachbin, a Brazilian mathematician who had been in turn a student of Laurent Schwartz. When Nachbin asked him if he wanted to go to Brazil in 1976, my father gladly accepted the offer.[6]

---

[3] My mom was never one to complain much, but I do recall a certain amount of bitterness when she talked about her college years: textbooks were both difficult to find, and often unaffordable, so she had to work off of her lecture notes as the sole course materials. The situation today has partially improved—textbooks are not as hard to find.

[4] Trinity College alumni include George Berkeley, Edmund Burke, Samuel Beckett, William Hamilton, Erwin Schrödinger, Jonathan Swift, Oscar Wilde, and many others.

[5] The restriction was counterintuitive: only after 1970 did the Catholic Church stop forbidding believers from attending Trinity College without special dispensation.

[6] Keen readers will note that the Mathematical Genealogy Project lists Gérard Emch as my father's PhD advisor. This is correct; things didn't work out in the end with Nachbin, and my father switched advisors halfway through graduate school.

Shortly after I was born, my parents decided to tear down the old house in Moravia where my mom had lived, and build a new house together. They timed their project just right: following some ill-advised monetary policy by then-president Rodrigo Carazo Odio, Costa Rica experienced a serious bout of inflation; the price of a sack of cement increased seven-fold during construction, forcing my parents to scale back their plans. By the end of the project, they were left with little disposable income. My mom often credited the *guayaba* and *cas* trees in the backyard for helping my parents get through the economic crisis, but I never understood exactly how.[7] I remember her being emotional when we had to cut down the *cas* tree in 1987; I was too young to understand the mixed feelings she felt towards it.

A picture of me as a baby being held by my mom, in 1981.

The struggles my parents experienced as they navigated the difficult economic turmoil of the 1980s were invisible to me. I was a happy kid; I loved to kick a ball around, though I was no good at it, and I collected Bazooka Bubble Gum tattoos. I entered kindergarten at the British School of Costa Rica in

(L) My parents when they met. (R) My father with my paternal grandparents in 1947.

---

[7] After showing my father a first draft of this *testimonio*, he explained: each night for some time, he and my mother would collect the windfall *cases* and *guayabas* and sell them to the local grocer. This would pay for the next day's bus fares to their workplaces.

Photo courtesy of Anthony Varilly-Alvarado.

A picture of most of the family in our kitchen. It was my mother's birthday, circa 1986.

1985, four years after the school began operating, and I graduated from it in 1998, having completed an International Baccalaureate (IB) Diploma. The British School is a small private school; my parents were determined to give my siblings and me the best education they could, and they wanted us to all grow up bilingual. Through most of my childhood and teenage years, my parents spent over half of their income on our education. It was a privilege and a sacrifice that we neither understood nor squandered. My mother had a few maxims she drilled into all of us. The most important one was perhaps *Papito, lo que tenga que hacer en esta vida, hágalo bien,*[8] but a close second was *Papito, cuando yo me muera no hay herencia, sólo su educación.*[9]

## Early Education

When I was in ninth grade, on a totally ordinary morning at school, my math teacher, Paul Murray, changed my life. Right before the beginning of first period, he intercepted me on my way to class, catching his breath, and asked me to follow him. He explained that that day was the first round of the Costa Rican Math Olympiad and that the school wanted to send five students from tenth grade to it. But one of the designated team members had not shown up. "Wanna go?" Being a goodie-two-shoes, I explained to him that I did not have permission to skip class. He rolled his eyes, pointed to a bus, and said, "Just get on the bus with the rest of the team, will ya? I'll take care of the rest."

I was good at math as a kid. Part of me felt like I had to be since my dad was a math professor at the University of Costa Rica. But I had never thought much of it. That morning, when I got to the first round of the Costa Rican Math Olympiad, something changed.

---

[8] "Whatever you have to do in this life, do it well." My father would usually abbreviate it this to "whatever you do, do it well," although his favorite version, following Robert Heinlein, is "anything worth doing is worth overdoing."

[9] "When I die, there is no inheritance, only your education."

I remember struggling for the first time. There were only 30 questions, for three hours, and they were multiple choice! But I had never seen problems like these. They required you to *think* in a way that school had never really challenged me. Some questions I could tell I simply didn't have the background for, but other questions were within my reach if I only spent time toying and struggling with them. It was exhilarating.

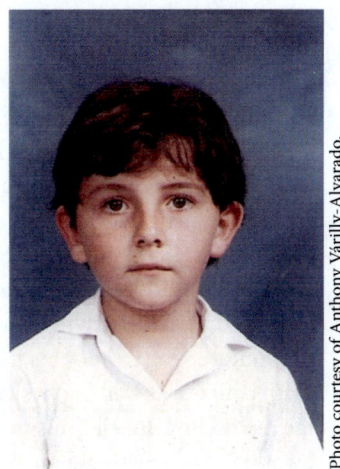

Throughout my childhood and adolescence, I was obsessed with astronomy. My contact with the subject came mostly from books at the University of Costa Rica's library and the odd purchase of *Astronomy* magazine, of which only a few copies could be found throughout the country, at ridiculous prices; I never owned a telescope. There was no visible professional community around the subject in Costa Rica, and this reality slowly eroded

At eight years old.

my dream of becoming an astronomer. I began drifting towards subjects more grounded in reality, like civil engineering, even though my heart was not in them. Math Olympiads, together with a role model and coach in my dad, rekindled my love of abstract subjects, of the pursuit of knowledge as its own end. I have been lucky and privileged in this regard, having been raised by supportive parents who nurtured and respected my hopes for the future.

International students in the United States often pay full tuition in college. Under ordinary circumstances, my parents did not make enough money to pay these kinds of sums, but by 1998, circumstances were not ordinary. My mother was being treated at M.D. Anderson Cancer Center, without U.S. health insurance. She would fly to Houston once a month for treatment. The hospital gave her many financial breaks, and the British School gave my brother and me a significant scholarship so we could keep attending school and finish out the IB program, but the financial strain on the family was immense. I had originally set my sights on studying abroad in England. But the fees for overseas students were simply out of my family's means. So I started looking into American universities, armed with AltaVista, the leading search engine of the day. It soon became clear that some universities awarded need-based financial aid, even to international students. All of them, however, were very difficult to get into: Harvard, Princeton, Yale, MIT, etc.

I'll never know exactly what the admissions officers at Harvard saw in my application. I filled it out by hand, wrote my college essay in 45 minutes, and in the page dedicated to summer activities (usually packed by most students with summer camps, internships at companies and research labs, sports and musical pursuits), I wrote "Not Applicable" and left the rest of the page blank. To be sure, I was a good student, very academically inclined, but so are many other thousands of applicants. I had written a small paper on triangle geometry for the extended essay component of the International Baccalaureate, which I attached to my application. Perhaps it helped? I thought I bombed the in-person interview, though I later came to realize that my interviewer, Renata Villers, a Harvard alum, must have gone to bat for me in a serious way.

# Higher Education

**College years**. I arrived in the United States on September 9, 1999, aged 19 and eager to study mathematics. My first semester at Harvard was rough. I felt like I was drinking out of a fire hydrant the whole time. I took Math 25: Honors Calculus and linear algebra, which wasn't the hardest class offered to freshmen (that was Math 55), and for the first time in my life I really struggled with math, but not in the "fun struggle" kind of way. I began doubting if math was for me after all. The professor, Kalle Karu, was phenomenal, so I figured *I* was the problem. I confessed my anxiety to my parents, who encouraged me to keep at it, to give my dream of studying math a chance until the end of the year. At the beginning of my sophomore year, my doubts hadn't gone away, so I began the year by taking some applied math courses, as well as physical chemistry. I loved those courses, but in the end, they made me realize how much I missed proof-based mathematics. By the fourth semester of college I was back as a math major, and I took abstract algebra from Barry Mazur and topology from Curt McMullen. Both were awe-inspiring instructors (and world-class mathematicians, though I didn't fully grasp this at the time). They rekindled my love for the subject.

Looking back at freshman year, I think the problem was one of time. I worked in the dish room of Annenberg Hall, the freshman dining hall, doing shifts on Monday, Wednesdays, and Thursdays, from 4:30 PM to 8:30 PM. By the time I got home, I was physically exhausted and had a hard time focusing on homework. Like many other international students and students from underrepresented backgrounds, I took on the campus job that paid the highest hourly wage. More savvy students took on library desk jobs, where they essentially got paid to do homework. I changed course my sophomore year when I became eligible to be a course assistant in the math department, though I lamented leaving Annenberg Hall; the staff there worked fantastically hard though their efforts seemed barely noticed by the surrounding students. But if I wanted to be able to rise to the same mathematical level as my peers who didn't have campus jobs, I realized I needed to take on a job that was more compatible with school work. I was fortunate to find such a job.

Harvard taught me a lot of things; it was a humbling experience, to say the least. Its pressure-cooker environment is not for the faint of heart, but meeting people who are

Photos courtesy of Anthony Várilly-Alvarado.

(L) Freshman year at Harvard. (R) Junior year at Harvard.

Photo courtesy of Anthony Várilly-Alvarado.

My family in Ireland, taken in 1994.

much smarter than you in what you consider your strongest suit is both disorienting and good for you. It made me realize that I liked to do mathematics because I loved the subject, not because I was decent at it. I learned that in the long run hard work will take you much further than innate talent. I also learned that meritocracy is a myth, having graduated in 2003 in the same class as Jared Kushner.

My mother died on July 8, 2002, the summer between my junior and senior years of college. That fall, my brother Patrick started college at MIT. Through sheer determination and will power, my mother beat the life expectancy cancer sentenced her to by more time than any of us thought possible. At her funeral, several of her co-workers remarked to me that she had lived to see her dream of making sure her children went off to college.

If I could go back and have a do-over, I would approach college much differently. I was hard-headed, and I almost never went to office hours (this was a *serious mistake*). I loved working with other people, which was great, but when the time came to apply for graduate school, there were very few professors who knew me well enough to write a strong letter of recommendation. I've now chaired the PhD admissions committee in the Rice Math Department for six years; so I know that letters of recommendation are probably the most important part of an application. You need letters from professors who know you well, who can speak to your potential for completing a good PhD thesis. It all worked out in the end for me, but this is part of where a bit of luck and the privilege of coming from a top-ranked school with good grades have played an important role in my life.

**Graduate school**. When I started graduate school at UC Berkeley in fall 2004, I had no idea what research was like in math. Research Experiences for Undergraduates (REUs) were not widespread in the early 2000s, and in any case international students were not eligible for National Science Foundation (NSF) stipends. Still, there were hurdles to be overcome before starting on research. For me, the hardest one was the advanced oral exam. I do well on written exams, but I freeze up on oral exams. To this day, when I am asked a question during a talk, I have to pause for a moment, breathe in and out a few times, and force myself to stop thinking that I am not thinking about an answer. My oral exam was rocky, to say the least. I passed it, though only because about an hour in I was certain I had failed, so I calmed down and was able to finally start thinking clearly.

Towards the end of my second year in graduate school, I proved my first research-level result. It was a small proposition, something I needed for a project. I remember the moment distinctly: I proved something that people didn't know already. It was a modest contribution to the sum total of human knowledge, but it was a contribution, and I was the author! That was the moment I finally believed *I can do this!* From there on out, I worked really hard on my thesis and a couple of side projects. Not everything panned out, and there were moments of intense frustration, anxiety, and anger at myself. I was extremely fortunate to have a strong group of friends in graduate school, including Dan Erman, Bianca Viray, and David Zureick-Brown; we supported each other through thick and thin. I also met some wonderful people who were a little older than me, at the postdoc stage, mostly at conferences, like Damiano Testa, Ronald van Luijk, and Mauricio Velasco. I learned a lot from them, and we collaborated in projects. There is no need to go at it alone. I learned this lesson serendipitously.

We do not choose our families. But we do have a lot of say on who we let mentor us. Mentors really matter. You need many of them: no one person has all the answers and all the advice that is appropriate at all stages of your life and career. I have had many people I gladly count as mentors. Among them, my thesis advisor, Bjorn Poonen, and my postdoctoral advisor, Brendan Hassett, really stand out. I knew I wanted to work with Bjorn ten minutes after I, as a prospective student, met him. I asked him about his research, and in order to answer me, he asked three questions back, to calibrate the state of my mathematical knowledge, without judgement. Once he knew what level to pitch his answer, I was blown away by what he told me. At the time, I didn't understand that I was drawn to Bjorn because he is a fantastic communicator. I love the area I work in, but I could have been happy doing many other things. Your relationship with an advisor is lifelong and is particularly intense during graduate school. Making sure the match between people is right is much more important than pursuing some specific subject. When picking a mentor, you should pick the person, not the subject.

One thing that helped keep me sane through graduate school was a stable personal life. Throughout graduate school, I lived with my girlfriend, Sarah, a wonderful, caring, ambitious and supportive human being, who helped me navigate the ups and downs of graduate school. There was a semester early on where we had to live off of my $1,200 month stipend in the Bay Area (rent was $950/month), while Sarah looked for a job. Although at our poorest, I remember those months as some of the happiest of our relationship. We got married in 2010, though as sometimes happens, we grew apart and divorced a few years later. This of course was a difficult time, but I was already a tenured professor, and my work could absorb the personal shock to the system.

**Postdoctoral years**. When I finished graduate school, I had some good offers for postdocs. One of them was at University of Georgia (UGA), another was at Rice. I happened to be visiting Atlanta while deciding where to go; a conversation with Danny Krashen helped me sort things out. Danny was trying to convince me to come to UGA. All I remember is that at one point he said, "OK, the truth is that there is no better mentor for you than Brendan Hassett [at Rice], but [...]" I honestly don't remember the end of the sentence. That moment crystallized things for me. I knew in my heart of hearts that Danny was right and that Brendan would be a great, if demanding, mentor. I took the job at Rice the next morning, on my way back to California. Years later I told Danny this story, and he gave out a big laugh. He told me something similar had happened to him earlier in life, and he was happy to have (inadvertently) paid it forward.

Some of Brendan's best advice came early on: "don't stall, keep moving," meaning prove the best results you can, but don't hold off until things are optimal before releasing them to the world, at least not when you are a postdoc. Also, be proactive in your search for new research problems to work on. The other piece of advice was "don't move away abruptly from what you know, work by analytic continuation," meaning take advantage of what you know already, and move towards where you want to be slowly, writing papers along the way. I can very easily trace a path between a paper I wrote on Zariski density of rational points on del Pezzo surfaces over number fields and a paper constructing pluri-canonical forms on moduli spaces of special cubic fourfolds (there are four papers in between). It'd be hard to find a conference with talks in these two topics. Related to this, don't spend six months learning an entire subject from the ground up because you might need it for a paper. Chances are you'll eventually realize the idea won't work, you'll have six fewer months to write papers before you apply for tenure-track jobs, and no new paper to show for your six months of work.

Brendan also taught me a lot about being a professional mathematician, about having long-term goals in mind, and problems of various sizes around those goals. I also learned a lot about *taste* from him (although this word was never used in our conversations). I learned that just because you have a hammer, and you see some nails in the distance, it may not be worth your time to go hammer those nails unless you have a good reason to do so. Our time is finite, and everyone succumbs to finitude. As you get older, you get more picky about the problems you work on, not because other problems are not interesting, but because each choice of a project closes doors on other choices. I cannot improve on David Foster Wallace on this point, so I will simply quote him wholesale:

> Day to day I have to make all sorts of choices about what is good and important and fun, and then I have to live with the forfeiture of all the other options those choices foreclose. And I'm starting to see how as time gains momentum my choices will narrow and their foreclosures multiply exponentially until I arrive at some point on some branch of life's sumptuous branching complexity at which I am finally locked in and stuck on one path and time speeds me through stages of stasis and atrophy and decay until I go down for the third time, [...] it seems unavoidable—if I want to be any kind of grownup, I have to make choices and regret foreclosures and try to live with them.[10]

My own outlook on life is nowhere near as gloomy as DFW's. I feel only gratitude for the privilege that I do something that I truly love that I get to share with students and colleagues. I have forfeited many other careers, some much more lucrative than my own. But I have no regrets about my choices. I once asked Ryan Hynd what he would do today

---

[10] This little reflection is embedded in *A supposedly fun thing I will never do again.*

if it were his last day on Earth. Without missing a heartbeat, he said, "same thing I had planned on doing this morning. And if that's not your answer, then what the hell are you doing with your life?"

## Tenure-Track and Beyond

In late 2011, I applied for tenure track jobs. There was a job opening at Rice, and even though it was a dream to stay there on a more permanent basis, it is highly unusual in mathematics for an institution to hire one of its postdocs into a tenure-track position. On a late Friday afternoon in mid-November, David Damanik, then the head of the appointments committee, knocked on my door and asked me if I was interested in interviewing for a tenure-track position at Rice. I'm not sure I kept my cool, but I immediately told him I'd love to.

"Good. How about Monday?"

This was a bit shocking and caught me unprepared. I mumbled something about teaching two classes on Monday.

"OK. Tuesday then?"

Realizing I could not delay the future much longer, I took the date. The interview went as well as it could have, though I didn't hear back about an offer until February. I was not the top choice for the job, and that's OK. I've never felt like I have a chip on my shoulder for that. Every year, hundreds of people apply for each available tenure-track position at a research-intensive university. I was under no illusions that I was somehow the department's top choice for a position, though I did feel like I could rise to the challenge of the job. The top choice candidate could only take one job; thankfully, they didn't want the job I dreamed of taking.

I earned tenure in 2016, and in 2019, ten years after setting foot at Rice as a newly minted PhD to take on a G. C. Evans instructorship, I was promoted from Associate Professor to Full Professor. Never in my wildest dreams did I imagine that that's what the future held in store for me when I first arrived in Houston.[11] Although I have worked tirelessly to get to where I am today, I recognize the luck and the privilege that have smoothed out my journey, and the sacrifices my forebearers made so that I could have opportunities to thrive. With a seat at the table, I now have the chance to help others thrive. I do not intend to waste the chance.

Today I live in Houston with my partner Carey, a smart, thoughtful, kind, and independent person I am happy to share my life

Photo courtesy of Anthony Várilly-Alvarado.

With my partner, Carey.

---

[11] As it happens, Houston already had a special place in my heart, even though I had never been to it: attentive readers will recall that my mother was treated at M.D. Anderson Cancer Center (without U.S. health insurance!) for the last few years of her life.

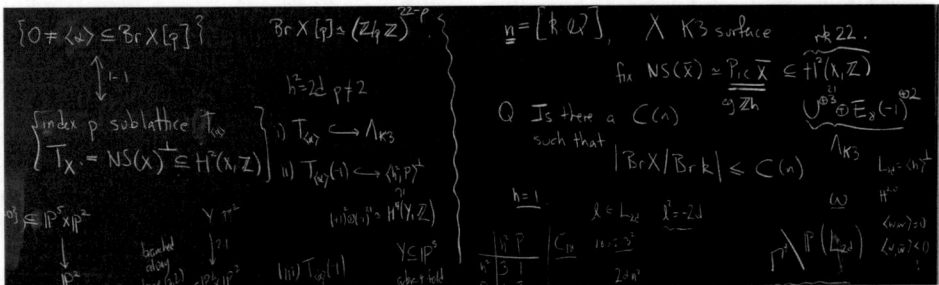

Chalkboard with a glimpse of my research. Photo courtesy of Anthony Várilly-Alvarado.

with. We have both suffered loss in our previous marriages, and this perspective, afforded by failure, pushes us to make sure we don't fall into the same traps of the past, or revisit mistakes. Before the COVID-19 pandemic we used to enjoy the arts and music scene in Houston, and we traveled together extensively. The pandemic brought a lot of our activities to a screeching halt, but we have found new hobbies together.

## Research

Most of my work to date is in an area called arithmetic geometry. I study Diophantine equations through a geometric lens. Perhaps the most famous Diophantine problem is *Fermat's Last Theorem*, which states that the only solutions to

$$x^n + y^n = z^n; \quad n \geq 3$$

are those for which $xyz = 0$ (i.e., at least one of $x$, $y$, or $z$ must be zero). A decade before Wiles gave his spectacular proof of this result, arithmetic geometers already had *good reasons* to believe that Fermat's Last Theorem is true: for each $n$, the Fermat equation defines an algebraic curve on the projective plane, and the general theory of curves already showed that, for each $n \geq 4$, there could be at most finitely many solutions to Fermat's equation. For a detailed explanation of how this is the case, I invite you to watch my talk at the 2020 Joint Mathematics Meeting titled *The Geometric Disposition of Diophantine Equations.*[12]

I've spent the better part of the last ten years studying equations that give rise to geometric objects called K3 surfaces. One of the most incredible surprises of my life has been a collaboration with people coming from electrical engineering, government security, and coding theory, to develop a geometric framework behind efficient systems for cloud storage, using K3 surfaces! I never expected that my knowledge reservoir on algebraic surfaces would be helpful in applications. Most projects I've worked on had applications internal only to mathematics when I began them, but now I've found that people from many kinds of applied fields can use these tools. I am currently working with earth scientists, using algebraic geometry to understand micro-layers of the earth's mantle from earthquake data! Theoretical mathematics has a lot to contribute to the world, but you have to understand it at a deep level in order to make the connections to the world around us.

I have the rare privilege of having a father who is also a mathematician. He works on non-commutative differential geometry. Our work within mathematics is quite distant,

---

[12] A link to the talk is provided here: youtube.com/watch?v=GnE2lFJ1x-Y.

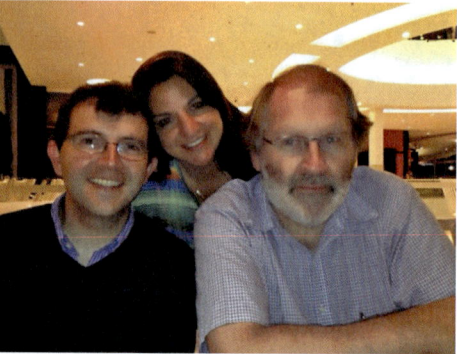

Photos courtesy of Anthony Várilly-Alvarado.

(L) With my Dad. (R) With my sister, Paola, and my dad.

but we share a common basic language that allows us to explain to each other our recent papers. In 2017, we finally converged in a conference, the Mathematical Congress of the Americas, in Montreal. And in 2020 he came to Denver to watch the lecture linked to above, which was an invited address of the American Mathematical Society at the Joint Math Meetings. I feel incredibly lucky to be able to share this connection with him.

## Conclusion and Advice

Some advice is peppered throughout the narrative above. Rather than rehash it all, I'll offer some further pointers for different stages of an academic career.

**Undergraduate years**. Less is more: fewer math classes, done excellently. This is a marathon, not a sprint. Early on, take proof-based linear algebra, abstract algebra, real analysis, and if possible a course in point-set topology (not all at the same time, of course!). Go to office hours. Connect meaningfully with your professors. Always ask *why*. To quote Ravi Vakil: "What is a group?" is not a great question. A better question is "Why is a group?" Whenever you meet a new definition, play with it through examples and ask yourself: why would anyone think it'd be a good idea to formally codify this concept?

**Graduate school**. The first year will be brutal. Hang in there. When you get over the hurdles of written and oral qualifying exams, your world will be turned inside out: up to this point, the pressure to get things done has been external, framed by structured coursework, homework, exams, etc. Now it's up to *you* to generate some internal fire to work on a research project. Think of it like a full-time job. Make sure you put in 40 hours of work per week. Above all, don't go through it alone: find a group of friends who can act as a support network. Read and criticize each others' work. When you look for an advisor, choose the person, not the subject.

**Postdoctoral years**. This is a hard, lonely time, with new mountains of responsibility. A good match with a postdoctoral mentor is key, but keep in mind that they are not your thesis advisor. Work hard, but take personal days regularly. Write all the papers you can, at the sweet spot intersection of interesting, feasible and within (or just beyond) your reach. Lean on your mentor to help choose projects.

**Tenure-track years.** Keep your eye on the ball. Minimize committee work. Keep writing papers. Enjoy the flow. Otherwise, why are you doing this?

**Tenure.** Keep moving. Time to give back. You will be busier than ever before, but in a good way. Mathematics requires community. You are now in a position to effect changes in that community, and by doing so, improve it. Do so.

Related to this last bit of advice, and perhaps most counterintuitively: be a bit selfish during your development as a mathematician. We Latinx people have strong family cultures, and often seek out opportunities to give back to a community that has given us so much. You will do so, in due course. First, get a seat at the table. From there you will be able to help more people than you ever dreamed of.

# 25

# Dr. Mariel Vázquez Melken

## Mexico City and the Early Years

I was born in the great Mexico City. Here the adjective *great* applies to many different things. First, the Mexico City Metropolitan Area is large and densely populated. People of all colors and creeds coexist within its boundaries. At the time of this writing its population is estimated at 21.7 million, and within the city limits the density is estimated at 15,600 residents per square mile.

Dr. Mariel Vázquez

Illustration created by Ana Valle.

In its second form of greatness the city has a rich history and is deeply cultural. According to a legend, the Mexica tribes traveled thousands of miles looking for the land where they were to settle. As they walked through a valley surrounded by mountains they found an island within the Lake Texcoco. There, perched on a cactus, was an eagle with a snake dangling from its beak. This was the sign. The settlement grew to become the capital of the Aztec empire, the largest city in the American continent. Tenochtitlán was founded in the fourteenth century by the Mexicas. When the Spanish conquistadors arrived it was a prosperous city with a sophisticated socio-economical system and urban architecture.

In his 1576 book *Historia verdadera de la conquista de la Nueva España,*[1] Bernal Díaz del Castillo wrote:

> "When we saw so many cities and villages built in the water and other great towns on dry land we were amazed and said that it was like the enchantments […] on account of the great towers […] and buildings rising from the water […]. And some of our soldiers even asked whether the things that we saw were not a dream? […] I do not know how to describe it, seeing things as we did that had never been heard of or seen before, not even dreamed about."

The Spanish conquered the city in 1521, and built Mexico City on the ruins of the great Tenochtitlán. Throughout the centuries, the lake was drained and the city expanded over it. Every time an earthquake hits Mexico City its citizens are reminded that they live atop the ancient lake bed. From one of my favorite spots in Mexico City one can witness

---

[1] *The True History of the Conquest of New Spain*

Photos courtesy of M. Vázquez Melken.

History of Mexico (clockwise from top left): Aztec calendar; Mayan *estela* (relief) from Yaxchilan, Chiapas; colonial house in the Coyoacán neighborhood in Mexico City; stucco facade of a Mayan temple from Placeres, Campeche. Three of the pieces are on display at the National Museum of Anthropology and History, in Mexico City. Credit: M. Vázquez Melken

500 years of history: the *Templo Mayor* (built between the fourteenth and the sixteenth centuries), the Metropolitan Cathedral (built between 1573 and 1813), and the *Torre Latinoamericana* (Latin-American tower, built in 1956). The photos above give a glimpse into the rich culture of Mexico.

The next measure of greatness is not as captivating. During my childhood the city regularly broke records levels of pollution and crime. Life amidst daily chaos was challenging. Home, family, and school were my protection, and schoolwork was the crutch I leaned on as I navigated the winding path from childhood, through puberty, and into adulthood. Outside the bubble there was turmoil. The 1980s started with one of the worst presidents in Mexican history. The ensuing years of financial instability were felt at all levels of society, with the rich getting richer, the middle class shrinking, and the poor falling in the depths of misery. We dealt with family tragedy, and lived through the earthquake of September 19, 1985. The 8.1 magnitude earthquake started at 7:19 AM and lasted almost

four minutes. We were used to the occasional movement of the earth, but this was different. Thankfully our house was not built on the lake bed, and it was not damaged. When
the earth stopped moving we tried to get over the fright and continued with our morning
routine; after all it was a school day and we could not be late! It wasn't until we got in the
car that the news of widespread damage reached us through the radio. Ten minutes into
our trip, we turned around and returned to our home. Schools were closed for many days.
I volunteered at the Red Cross, was given gloves and asked to go through a mountain of
debris, in search of lost personal belongings. At least I felt that I was doing something to
help, and millions of people came together as the government struggled with their response. *This experience taught me early on about the power of grassroots efforts and the need
to have empathy, believe in our community and cater to it.*

## Family

In the face of adversity we always turned to family unity and to education. Education is
life insurance. Instead of favoring a fancy house, or a move to a neighborhood that would
cut down on our daily commute, instead of indulging in expensive vacations, my parents
prioritized education. They chose to pay for tuition in a school that was to provide me
with the foundation needed to succeed in my later studies. Still, in the social context of
our country, we were among the privileged. I understood this and have never taken it for
granted.

In the first half of the 1980s, when the crisis hit the hardest, our vacations consisted of
day trips to the many small communities around the city, and bi-yearly trips to see our
grandparents in Mérida and Campeche, two beautiful cities in the Yucatán peninsula. The
road trip lasted between 16.5 and 22 hours, depending on traffic and on road conditions.
My dad did not like to stop and often chose to do the trip in one go. I dreaded the gas
station bathrooms, and got nauseous in the car. So I slept, as much as I could. Once we
arrived to our destination our wonderful grandparents awaited us along with crates of
mangoes and other tropical fruits, coconut water, the smell of the tropics, and the movement of sunlight on the water. It felt like paradise, and we immediately forgot the nuisance
of the trip. It was worth it. I learned the value of family, and that personal sacrifice is out-
weighed by giving joy to others. As tweens and teenagers it would have been easy to whine
our way into staying home. That would have been a tremendous loss.

## Discovering Mathematics

I had a happy childhood, surrounded by a strong and loving family. I remember fondly
the frequent weekend outings to the city's parks or one of the forests in the outskirts of
Mexico City. As a little child I loved counting and finding patterns. Every time I had a set
of items at hand, I sorted them into similar shapes and colors, and I counted them. I also
loved to see geometrical patterns around me. These ranged from daily occurrences on
tiled floors and walls, to the painted or embroidered patterns on the artisan work that we
saw during our trips. I admired those extraordinary geometrical shapes carved into the
stone temples of the Aztec, Mayan, Olmec, Toltec, and Zapotec cultures. I also saw patterns in nature, twisted twigs, entangled seaweed, the intertwining of the waves, and the
interplay of light and movement during a sunset on the water. To me, math went beyond

numbers. It also consisted of shapes, colors and movement, and it was partly art. As the years passed, my love for mathematics grew. Mathematics was everywhere, but the idea of devoting my life to it, was quite abstract. I did not know that one could do math for a living, and thus I assumed that I would become an engineer, like my father and grandfather.

At school, I worked extremely hard and started building my vision for the future, however uncertain it felt. In high school, I focused on math and science. The discovery that I could become a mathematician made me very happy. In my senior year I learned about DNA and fell in love with molecular biology. As the time to apply for college approached, I perceived the private universities as too narrow and limiting in their offerings. I had a thirst of knowledge and it soon became clear that my ideal school was the national university. The math curriculum was flexible and enthralling, most courses looked fascinating. Studying math seemed to necessitate giving up biology. I convinced myself that I could become a mathematician and later pursue a master's degree in molecular biology. *Although the plan was unconventional, I persisted. I always followed my dreams.*

My transition from high school to college was smooth. The National Autonomous University of Mexico (UNAM) offered me a first-class education. And it was free of cost. The transition from the tight bubble that enveloped me through high school to a university system that, in 1990, enrolled more than 271,000 students, employed 28,000 academic personnel and 25,600 staff members, may have seemed daunting. I thrived. I flourished in the immensity, the beauty, and the cultural offerings on a campus fittingly called *Ciudad Universitaria.*[2] Each day, the sensation as I entered the campus was that of coming home.

## La UNAM

In Mexico, university students live with their parents whenever possible. Those who have to move away from their hometown to attend college live with extended family, primos, tíos, abuelitos.[3] In the absence of extended family, students, especially women, lived in vetted and 'respectable' guesthouses. The university had no dorms, and the rental market was intimidating and not affordable. Most also perceived living alone as dangerous, and a sign that you were devoid of your family's protection.

I stayed home, with the caveat that my home was far from the university, even by Mexico City standards. I lived in the northwest of the metropolitan area, while the university was located in the southwest. Distance is relative in Mexico. One can measure the absolute distance, which in this case was a mere 25 km (approximately 15 miles). The temporal distance was more practically relevant, and it varied from 25 minutes in the wee hours of the night (2 AM, for example), to 45 minutes if you were lucky and your commute was light, and up to two hours on the worst of normal days. Abnormal circumstances included road flooding after a storm or road blockage due to a crash, construction or one of the frequent marches for social justice. On those days you had better stay put and wait until late at night to attempt the commute back home. Of course, the temporal measure of distance had to be weighed by the means of transport. I was fortunate enough to own an old car that I could use for my daily trek to University City. Most citizens did not have that luck. Commuting by public transportation meant taking one or two small buses (the

---

[2] University City
[3] cousins, uncles/aunts, grandparents

infamous *peceros*) to the nearest subway station, transferring twice, and finally arriving at the University City Station which was a ten-minute walk from the math department. All in all this was an exhausting 2.5-hour one-way commute. When driving I had to leave my home at 7:30 AM to make sure that I made it to my 9 AM class. As the years went by and traffic worsened, I left earlier and earlier. Leaving at 6:45 AM would cut the traffic in half allowing me to cross the city in 45 minutes, arrive early and study on campus, or take the occasional 8 AM class. My preferred commute was a hybrid: leave home at 7:15 AM, drive for 30–40 minutes to *Colonia Roma*, exercise, grab a coffee and take the subway into campus.

Occasionally, I commuted with a friend. It was good to have someone to talk to while braving the road rage and the traffic in Mexico City. He once told me that if I didn't become a mathematician I would surely find a job as a cab driver. I learned every possible route and shortcut on my way to school. My mission was to not sit in traffic, so I often ventured off the main road, and developed a useful skill. It sounds funny, of course, but underlying it was a sinister cause: pollution and road chaos get on your nerves. They generate mounting anxiety. A few times I felt on the verge of breakdown. Optimizing my trajectories from point $A$ to point $B$ in the city entertained my brain and kept it from going to dark places. Unbeknownst to me, this presaged my future interests in random polygons. After all, each trajectory from home to college and back was a polygon in three dimensional space, whose embedding was affected by the fourth temporal dimension (time of day, day of the week, etc.) and by the randomness conferred to it by the city itself. Observing the world around us, and navigating it, teaches us about shapes and about the dynamics and randomness of different processes. *Even adverse situations offer opportunities for reflection and learning.*

Around that time I found a deep love for theoretical mathematics. My favorite classes were those in topology, geometry, set theory, number theory and graph theory. I started attending the national meeting of the Mexican Mathematical Society, and the Graph Theory Colloquium, soon yielding to the prospect of becoming a researcher. I am grateful to so many wonderful teachers, to my classmates, to our philosophical musings and love for math, and to the flexibility of the system.

At UNAM there were no majors or minors. The college application involved a general and a topics entrance exam. I forget the exact sequence of events, but I do remember having to indicate my chosen subject, as well as second and third choices. The first choice was mathematics. The second, in case you are curious, was architecture. Admittance implied entry into the *carrera*[4] of mathematics at *La Facultad*.[5] The *Facultad* consisted of a cluster of buildings and subjects that included mathematics, physics, computer science, actuarial sciences, statistics, and biology. All courses were chosen from an extensive list of mathematics courses provided by the math department. The first two years were quite rigorous and predetermined, but after that the flexibility was exhilarating.

The memorable Calculus II classes from the power team, Luis Briceño and Julieta Verdugo, and Calculus IV from Javier Paéz, not only gave me strong foundations, but also taught me how to teach. To this day I find myself shaping my lectures inspired by

---

[4] career
[5] the Faculty of Sciences

theirs, with lots of in-class discussion, long homework sets, and by integrating a research project into the student assessment. Briceño and Verdugo demonstrated team science in the classroom, were fantastic lecturers, and gave students a glimpse into the work they did outside the classroom. Several times a year they offered training sessions for K–12 teachers in low-income public schools—an essential service to a population ravaged by poverty, decades of underfunding and poor teacher preparation. I remember my teachers and I am grateful for the education they delivered with passion and dedication. *A fantastic teacher can change an undergraduate's life and start shaping the mold of a future researcher and mathematician.* Why then do some of our institutions, and colleagues, look down on those professors who are excellent educators, but do not do much research in mathematics? I personally think that there is room for everyone and that higher-education benefits from a diversity of approaches to teaching. Each instructor meets their students at a different level. Some will capture the student's imagination in high school or in freshman year. For example, I learned introductory astronomy from Julieta Fierro, who on the first day of the semester, entered the classroom in roller blades, and climbed on the table to illustrate heaven in the Babylonian cosmogony. Dr. Fierro inspired a lifelong fascination for the skies and a deep appreciation for science communication. In mathematics some instructors can inspire the jump into deeper and more abstract mathematics through the calculus or linear algebra series, when a student could as easily switch to a different, not so challenging, discipline or major. Other professors can shape students dreams early on by modeling the path into abstraction and the tenacity needed to go to graduate school in mathematics and to undertake mathematical research.

Did I mention that UNAM was an oasis amid the surrounding chaos? The comings and goings of mathematicians and other scientists modeled a more civilized society where citizens look out for each other, and work hard for the sake of knowledge, with no financial interest, with the goal of educating the next generation. Our facilities were not fancy, but we had all that we needed. Education was free and all services, including books, photocopies and food were heavily subsidized. I am not familiar with UNAM's budget model of the 1990s, but from the point of view of the student I can tell you that we had what we needed: good professors, high-quality curriculum, large classrooms (clean and with chalk on the boards), green areas, a small library and a cafeteria. Minutes away was a cluster of world-class science institutes (astronomy, geophysics, materials science, mathematics, applied mathematics) with access to top-notch researchers, and the first and largest super computer in North America. A short drive, or shuttle ride, away was the massive central library, and the school of medicine. Going southwest were the volcanic formations and the university cultural center where I saw more symphony concerts and watched more art movies than I can count; excellent offerings at a very low cost for students. UNAM catapulted me into the possibilities of my future and I am ever so grateful for that.

## Knots and DNA: The Launch of My Career

After taking the core courses I started leaning towards the theoretical math offerings. I took two number theory classes from the legendary Alberto Barajas, one of the founders of modern mathematics in Mexico. I learned graph theory from Neumann Lara, a wide range of the topology and geometry offerings from Bracho, Clapp, Montejano, Eudave,

*Licenciatura en Matemáticas*, bachelor's degree in mathematics, from the National Autonomous University of Mexico (UNAM). This was a tremendous source of pride for my family, the culmination of many hard years of work and the constant reminder that not everyone is as fortunate and as privileged. The university motto, *Por mi raza hablará el espíritu* (Roughly translates to: through my people, the spirit will manifest) is a guiding force that maintains that the way forward is through education.

Gómez Larrañaga, González Acuña, Neumann Coto, four semesters of analysis from Grabinsky and Carrillo, set theory from Amor. Back then I thought of "pure" math in opposition to "applied math" and inferred that I would need to relegate my interest in molecular biology to a mere hobby. I was mistaken. One day, walking through the halls of the department, I found a poster announcing a series of lectures on "knots and DNA" by De Witt Sumners, professor of mathematics at Florida State University. This combined the math that I liked with molecular biology. I have thought about knots and DNA since that day. It has been the leitmotif of my career. This day led me to the path less traveled and shaped my future. *Work as hard as you can and follow your dreams as they will take you exactly where you need to go, even when the path may seem unconventional.*

An undergraduate thesis is a graduation requirement for college students at UNAM (and most other universities in Mexico). I have often claimed that a third of what I learned in college I learned while writing this thesis. In bulk amount of material the thesis can probably not even come close to the hundreds of derivatives, limits, and integrals solved in the calculus series, but if we consider the content weighted by its future impact, the thesis largely overtakes the rest. That being said, doing research would not have been possible without the foundation acquired during the first few semesters of my college years. Beyond enjoying the details of the work itself, writing a thesis moved me into the world of mathematical research. I became an undergraduate scholar in the math institute, attended national and international conferences, listened to famous researchers and witnessed the lifestyle of professional mathematicians.

It soon became clear that I wanted to go to graduate school and pursue a PhD program. However, I had not asked my parents to pay for a college degree, and was surely not going to ask them to come up with tens of thousands of dollars to pay for graduate school. The idea of applying to a PhD program abroad did not materialize until I understood that I would not only not be expected to pay for the degree and support myself, but that the university that admitted me would pay me a student salary sufficient to support my living expenses, and would cover my tuition. I always wonder how many people don't even start dreaming because of the fear of the financial cost. Those individuals could have brilliant careers and instead leave the pipeline.

The community supported me, and encouraged me. I received a scholarship and was admitted into a handful of prestigious universities. The choice was now easy. A few months later I boarded a plane with two suitcases and U.S. $500 in my pocket. My savings vanished in the first week of paying for necessities and various utility and housing deposits. But I was there, I was not afraid and was determined to succeed. I also had a fantastic advisor, De Witt Sumners, whose family helped me tremendously during the first few weeks.

## On Being a Woman in Mathematics

I am an optimist by nature, and tend to be very positive. I feel compelled, however, to unveil some of my experiences with injustice and misogyny. These are of course not limited to my city, or to the developing world for that matter, but I did not know it when I was young. We always think that the grass is greener on the other side. When I lived in Mexico it was impossible for a woman to walk in the street without being catcalled. For teenagers or young adults, the verbal attacks were vicious and continuous. On occasion they transcended the verbal and the perpetrator followed you for a few blocks on foot or for miles by car, adding to your fear, the constant terror of sexual violence. At secondary school I felt safe. In college I learned to be vigilant as I traversed the city from day to day. While on campus I felt mostly safe. I learned to avoid sensitive areas and to speed up my walking from the subway to the department. Inside the *Facultad de Ciencias*,[6] life was good. Well, it was good until I started working as an undergraduate research assistant, attracting the attention of one too many middle-aged professors. I was very shy and kept to myself, but I was driven and a very good student. This, compounded by the oddity of

---

[6] College of Science

a woman in mathematics, appealed to certain types. Of my freshman class of eighty, ten graduated in mathematics, with only two women. I was curious, focused and passionate about learning and breaking barriers. They also found these traits appealing. I was friendly and always carried a gentle smile. This opened the door to abuse of power. The few women doing research, most of them young students, learned to assimilate in the male-dominated world and to live with constant microaggressions in the form of false praise that some took as compliments, some relished, and most others dreaded. And then there was the joking… Many sexual or misogynistic jokes were (and still are) socially accepted. Making fun at the expense of women, minorities, and those with different sexual orientations was normalized. Oh, but do not take me wrong, this was not unique to Mexican society and the Mexican academic environment. I continued my career in the United States and the jokes continue until this day. They are less loud and are concealed under a veil of hypocrisy. There was for example the warning from a fellow graduate student to be aware that professor $X$ just stared at your breasts while he pretended to listen to you. The misogyny in Mexico was overt and accepted, while in the United States it is closeted, but omnipresent.

It wasn't until several years into my move to the United States that I learned to recognize discrimination. I have to say that such naïveté and ignorance was helpful. Looking back I now remember those instances and recognize them. For example, a staff member who, throughout the PhD, confused me with the only other Latina in the program. Our names are vastly different and each of us has a distinctive physiognomy, including different skin color. Or that famous mathematician who, after seeing my husband pushing our first child in the stroller, told him that his life as a mathematician was over. We were both postdocs dealing with the uncertainty of the future and navigating our first year as parents while teaching and conducting research.

These were signs of systemic gender and race discrimination, but I always chose not to linger on them. I thought that the perpetrators were individuals (as opposed to groups) making bad choices, and I tended to give people the benefit of the doubt. I still do, and I am the better for it. My approach has been to distance myself from them. Ever the empath, I rationalize what could have caused the behavior. I do not (or try not to) take it personally, brush it aside and keep moving forward. *We make choices in life and the choice to allow the attacks of others to hurt us is one that shapes who we are.* There is adversity in life. We cannot know when and where it will strike, but we can choose how much we will allow it to weigh us down. This, I recognize, is easier for those of us who grew up abroad, nevertheless, *we have a responsibility to recognize injustices, step up to protect others, and continue moving forward, one step at a time.*

## Conclusion

Today, fast-forward two decades, I am a full professor in the departments of mathematics and of microbiology and molecular genetics at the University of California Davis. I am married to my collaborator, life partner and best friend, Javier Arsuaga, and we have two beautiful children. I devote my research life to studying the molecular intricacies of genomes and the inner working of proteins that interact and change the topology and geometry of DNA. I co-lead a research group with my research and life partner. I find much joy in mentoring students in research and in communicating science at all levels. The other half of my academic life is devoted to combating inequities and providing an equal playing ground for all. I lead the Center for Multicultural Perspectives on Science (CAMPOS) whose mission is to support the discovery of knowledge by promoting women and other groups underrepresented in STEM, through building an inclusive environment. My academic and personal lives intertwine like the DNA double-helix. For now, while still young, our children follow their parents in the uncertain winding path through life. There have been many hurdles, but it is always about looking up and moving forward, with patience and concentration, putting one foot in front of the other. This is a story for another day.

# 26

# Dr. William Yslas Vélez

## Marriage and Partnership

Bernice and I met when we were 18 years old.
We were children, naïve. Our backgrounds
were surprisingly similar. We both had families
in Magdelena de Kino, Sonora, Mexico, and as
children, we both spent part of the summers in
that town. On our first date, we went to a wedding
shower at El Rio ballroom, where we danced *boleros*, *corridos*, and *cumbias*.[1]

Illustration created by Ana Valle.

Dr. William "Bill" Vélez

We married when I came home from my first
tour to Vietnam in the U.S. Navy. I was at sea for
most of our first married year, and this made for an
extended honeymoon. I could not write about my
life as a mathematician without also describing the
joy and support that I received from Bernice and
the family. Entering the mathematical community
could have been like entering a not-too-friendly country, whose language I could barely
speak. Instead, I walked in with an arsenal of support.

Every married couple goes through hardships and ours was no different. With large
extended families, there are many personalities and viewpoints. The children adopt
behaviors of the dominant culture, which have to be begrudgingly accepted. However,
we overcame problems and walked through life as lovers and partners. Bernice brought
complementary skills to our marriage, which made my life more beautiful, more meaningful, and more delicious. We now spend our time together cooking, still listening to the
Mexican music of our youth.

## Bill's Early Life

I was born in 1947 in Tucson, Arizona, and I grew up in the loving embrace of the
Mexican-American community. I found out much later that we were poor, just like the

---

[1] *Boleros* are a genre of slow-tempo Latin music. *Corridos* are narrative songs and poetry that
form ballads. *Cumbia* is a musical genre that originated among Afro-Colombian populations and
later was popularized throughout Latin America and the U.S.

259

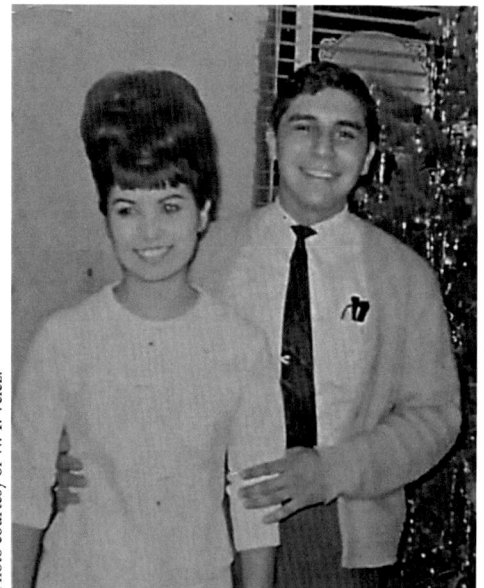

Bill and Bernice at her Nana's house.

rest of the community that surrounded us. I did not internalize this poverty, surrounded as I was by the richness of Mexican culture. I was well cared for and loved. More than anything else, I was well educated. In my home, a song of education played constantly. We were the only family in the neighborhood that valued books. We had a set of encyclopedias and another set of books on nature.

My mother, who was very proud of the fact that she graduated from Tucson High School despite the then-current attitude that women did not need such an education, would often say to us, "*Lo único que les puedo dejar es una buena educación.*"[2] My two brothers, Manuel and Gilberto, and I internalized this inheritance and we all graduated from the University of Arizona (UA). Though we led very different lives, we all passed on this song of education to others as teachers of English, music, and mathematics.

My parents were both born in Sonora, Mexico. My father, Emilio, was born in 1908 in Magdalena de Kino, and my mother, Julia Yslas, was born in 1910 in San Miguel de Horcasitas. At that time, the border between the U.S. and Mexico was fluid and people moved back and forth between Sonora and Arizona. There were families on both sides of the border and children would be sent to live with relatives on the other side of the border for extended periods. As a child, I would spend a good part of the summers living in Magdalena de Kino, and I had the great pleasure of mingling with cousins from both sides of the family. Before my father died, we would travel to Magdalena de Kino about twice per month. The ritual was always the same. We would first arrive at my paternal grandparents' home where we were received with love and affection, always arriving with gifts of food and other goods. My father would stay with his family and my mother and the children would then drive on about five blocks to stay at my maternal aunt's home, which was part of their hotel, *El Cuervo*. As children, we would roam the small town of Magdalena de Kino freely, visiting with many different family members. It was an idyllic life.

We were wild children. Life was safe and we could roam freely. Living in the Arizona-Sonora desert gave us a world to explore on our bicycles. An empty lot in front of our house was the gathering place for the kids. We would have neighborhood water fights with an arsenal of weapons. We played kick-the-can and hide-and-seek. In the evenings, adults would sit on their porches to try to catch some breeze. Food was enticing. Is there anything better than a fresh homemade tortilla with butter? My mother had a hard time keeping up with us as we passed by the kitchen to pick off a freshly made tortilla.

On our trips to Magdalena de Kino, we would bring back a variety of firecrackers. A *palomita*, a triangular-shaped firecracker, four inches on each side, placed inside a mailbox

---

[2] My inheritance to you will be a good education

would rip out the rivets and open it up for airmail. We found Josefina's (my sister) baby buggy in the open lot (or maybe one of us placed it there). This reminded us of covered wagons going across the prairie. We attacked the covered wagon with flaming arrows. Several explosions were heard when the stagecoach caught on fire. This confirmed our suspicions that the wagon train was part of a cavalry resupply unit. No dolls were harmed in this egregious incursion into our sacred empty lot.

Bill and Bernice at his mother's house.

Photo courtesy of W. Y. Velez.

## Growing up in Poverty

Growing up, we lived in a house that my father had built. My father, a mechanic, also had a garage built adjacent to our home. When I was about seven, our family went through financial hardships. My father's garage was sold to a good friend, Cristóbal Redondo. In order to save our home, most of our home was rented out to a family. We kept two rooms of the house, plus a small trailer that had one bed, where my father had always slept. We had no running water in the two rooms, no heating, and no cooling. We used the bathroom and shower in the garage next door. One room served as a bedroom and the other room as the kitchen and dining room. There were six of us, my parents, my three siblings, and myself. Next to our two rooms was a concrete floor, perhaps 10 feet by 12 feet and over it were grape vines. This concrete floor was separated from the adjacent street by a three-foot-tall wire fence. When the weather permitted, we slept on cots under the grape vines. The sun was our alarm clock. Despite this poverty, my siblings Josefina and Manuel were sent to piano lessons every week. I never learned to play a musical instrument and Gilberto was fortunate that a children's mariachi group was started in his school and he learned to play the string instruments of a mariachi group.

When I was in seventh and eighth grades, home dance parties were very popular. For me, at that age, there was nothing more wonderful than holding a girl in my arms and swaying to the music. At home parties, there were no teachers around to monitor the rule of maintaining a distance between dancing partners. Since kindergarten, there was always a girl I was attracted to and these parties gave me an opportunity to dance with that girl.

I mention these dance parties as an indication of my ignorance about our poverty. How could I have asked my friends to come to our home? I couldn't invite them in. There was no room. My guests had to sit in this small outdoor floor, which also acted as our bedroom in summer. Worse, what if someone had to go to the bathroom? They would have to go to the garage next door. Apparently, I was blind to the poverty that I lived in. My life was so rich with music, with freedom and wonderful Mexican food.

Bill in 1968.

After selling the garage, my father rented a gasoline station from my maternal uncle, Augustín Islas. I started to work there when I was eight, pumping gas and fixing flat tires. My father died when I was nine. My brother Manuel, who was thirteen, took over the management of the gasoline station and was responsible for its day-to-day management. He hired one of our maternal uncles, Francisco Islas, to work at the station during the day, and after school, Manuel would take over and close the station at 9 PM. Manuel would buy old cars, hire a mechanic to fix them and then sell them. He was always looking into how to bring in a bit more money.

When I was a bit older, I began taking over the late afternoon shift and would work until 9 PM. During the summers, I would walk the half mile to the gasoline station to open it at 7 AM and I would work until 9 PM. My pockets served as the cash register, right-hand pocket for coins, left-hand pocket for bills to make change. There were no credit cards then. Gasoline sold for 17 cents per gallon. Teenagers would show up with their 22 cents to buy gas for the evening. The business was slow. Fortunately for me, I was a voracious reader and passed the time reading books. In the seventh or eighth grade, I submitted over 120 book reports.

My mother was serious about education. I have no idea how she managed it financially. All of my brothers and I attended Catholic schools, first All Saints Catholic School until eighth grade, then Salpointe Catholic High School. She worked three jobs. One of those was selling Stanley Home Products. She would arrange a demonstration party at someone's home, display goods, and take orders. Stanley Home Products would send my mother large quantities of three-cigarette packages to hand out at these demonstrations. Cavalier cigarettes were the brand. All of the neighborhood kids smoked Cavalier cigarettes.

## On Mexican American Culture

I grew up in a very diverse environment. We lived at the corner of 34th Street and 7th Avenue. On 34th Street, between 6th Avenue and 8th Avenue, there were at least three Anglo[3] families, one Yaqui family and another Tohono O'Odham,[4] but the majority were Mexican American. At All Saints School, at least a third of the students in my class were Mexican American and there was one African American student. In grade school, all of

---

[3] One of the ways Mexican Americans referred to white Americans at the time.

[4] Both Yaqui and Tohono O'Odham are Native American people of the Sonoran Desert, residing primarily in the U.S. state of Arizona and the Mexican state of Sonora.

my teachers, except for one, were nuns. Perhaps it was this Catholic environment that made overt racism a rarity. However, high school was different. There were still many Mexican Americans there, but I felt that the school was trying to make me into a "white boy." I reacted against this by more fiercely embracing Mexican culture and adding my mother's maiden name, Yslas, to my name.

Growing up we understood that we were second-class citizens. The movies portrayed Mexicans as bandits. The movie about the Alamo had a profound impact on me. The white North Americans were the heroes, and the Mexicans were the bad guys. However, all of my family were Mexicans, and I could not internalize that message. I rebelled against this view of ourselves, and I reacted with a racist attitude towards white dominant North American culture and people. They didn't want me and I didn't want them. I was an angry young man, and my mother worried about the crowd that I was hanging around with. "*Dime con quien andas y te digo quien eres.*"[5]

Not only were Mexican Americans looked down upon, but even within our culture, there was discrimination. We were darker and more Indian-looking than many of our cousins.[6] My paternal grandmother was fair-skinned and had blue eyes as do several of my cousins. Women at that time were told to wear sun hats so that they would not become darker. Added to this, we were poor and had no father. My mother worked three jobs to maintain a home while all of my aunts were stay-at-home moms, as was the norm then. She was often criticized for not being home.

Many minority youth experience alienation from the dominant culture and, like me, react by not wanting to participate in a culture that demeans them and does not value their background. I never had a Mexican American teacher through high school. In my undergraduate and graduate studies, I remember only one Mexican American teacher, and that in a humanities class. What did that say about the relevance of education for my community? Over the decades, we have lost so many talented minority students because they could not see themselves on a path leading to college. These students saw too many exit signs along that path, and for many, their high school experience simply obliterated the roads that led to further education.

How did my mother do it? She provided my brothers and me with the best possible education available in Tucson. The financial hardships were significant in sending us to Catholic schools, but what is even more challenging was that somehow she guided us into this educational pathway. We all knew that she worked hard outside the house, but we also knew that she would stay up late into the night washing our clothes and ironing our Levi's jeans. As I grew up and began to loosen the anger that I felt towards Anglo society, I also gained respect for my mother, my family, and for our culture. This respect began to impact my actions. What would my mother say if she saw me doing this? That thought guided me. I was proud of our family and the respect that they had for education. This pride was an important tool for me that provided the extra impetus to forge ahead in times of difficulties. As I wrote in the dedication to my PhD thesis, "… *si me he avanzado, es porque soy Yslas-Vélez. Arriba la familia.*"[7]

---

[5] Show me who you are with and I will tell you who you are.

[6] My recent Ancestry results indicate that I am 25% Native American.

[7] "If I have succeeded, it is because I am Yslas-Vélez. Long live family."

# Higher Education

I did not display any particular talents in high school. I was never the best student in a class. I thought I was very well prepared for college though I had not the slightest idea what career to follow. I have written about my four-year career in college and I only want to mention how our wedding occurred.

We planned to marry in May 1968, just after graduation. By February 1968 plans were well on the way for this wonderful event, but then the U.S. Navy intervened. I was ordered to report to Long Beach, CA on March 27, 1968 to be processed for sea duty. By April, I was on board the USS Yorktown in the Tonkin Gulf. Fortunately, I had enough units to graduate, but all wedding plans were cancelled, invitations thrown away. In June 1968, President Nixon had a troop withdrawal of 5,000 troops and the Yorktown was ordered back to homeport. I called Bernice by short-wave radio as we sailed back to Long Beach. She had three weeks to plan a wedding. Family and friends jumped in to prepare the traditional wedding food: *birria* (a prepared beef), beans, rice and tortillas, enough to feed over 400 guests. I arrived in Tucson the night before the wedding, we were married and that night we began our return to Long Beach, but I was soon out to sea again.

In mentioning the following set of events to a mathematician, he remarked that I should have been the poster child for failure. I started graduate school in 1970, having just returned from Vietnam serving on aircraft carriers. It had been two years since I had completed my undergraduate degree. I earned many Cs in advanced math courses, I was married and our first child, Ana Cristina, was born in the third week of that first semester. I was a teaching assistant, teaching two courses per semester. The salary of $3250 per year was not enough to live on, but I also had the GI bill, which provided educational assistance for veterans and service members. I was fortunate that my brother-in-law, Francisco Redondo, would have side jobs on some weekends, and I would work as a laborer for his masonry crew.

Not only did I not fail, I thrived. The five years in graduate school were among the happiest in my life. I was learning a great deal of mathematics, I was married, I had two children to come home to, (Andrés Antonio was born on March 25, 1974), and I owned a house. When I returned from Vietnam, I returned with a little bit of money (you don't spend much out at sea). Bernice, who had taken over our finances, suggested that instead of paying $75 per month in rent, we should look for a house to buy, which we did, and the payments were $135 per month. It was a fix-up special. The week after we

Photo courtesy of W. Y. Velez.

Bill and Bernice, July 20, 1968.

Photo courtesy of W. Y. Velez.

Building a playhouse for Ana, 1973.

bought it, my two brothers and I put up a new roof. I also used my masonry skills and built brick walls to line our flower gardens and I also learned how to do basic carpentry.

In March 1975, I had completed my thesis and as a gift to myself I bought a skillsaw and a sander. I built a living room set out of pine lumber: sofa, love seat, and end tables. This set was indestructible, and I called it my first approximation to a couch. Later I found springs from an old couch and created the second approximation. After 15 years or so, Bernice got tired of it and I came home one afternoon to find out that she had sold the set for $200.

## Overcoming Naysayers

I have often heard that being married in graduate school and having children makes graduate school much harder. I found it to be the opposite. On the day that our daughter was born, I did not turn in a good homework set for my algebra class. The next class day the instructor walked into the classroom and handed back the homework. When he came to mine, he wadded it up in a ball, threw it at me across the room and yelled, "You will never be a mathematician." Such an incident can be devastating to a person, but I went home that night and held a baby in my arms. I felt that my family was a place of refuge. Later, that same professor looked at one of my solutions and asked me to take his course in algebraic number theory the following semester and to think of working with him. He became my thesis advisor.

I took his course in number theory and I did agree to work with him. I became a number theorist, yet I only took a one-semester course in number theory; I did take supporting courses in group theory and function fields though. Henry Mann, my thesis advisor, took a real interest in me. I would talk to him regularly and in the course of our conversations, I would discuss my family situation. I was born in Tucson, half of Tucson was related to me. Every weekend there was a baptism, birthday party, a *quinceañera,*[8] or a wedding; and every other Thursday a funeral. I was also the family plumber. Besides my family commitments, I taught two classes per semester for the five years that I was a teaching assistant. I was pressed for time.

Henry Mann had us over for dinner one evening in my second year. As we drove home that night, Bernice mentioned that Mrs. Mann had taken her aside for a talk. She told Bernice that if she wanted me to be good, she had to learn to leave me alone. Bernice was quiet for a bit and then her quiet, confident voice came out of the darkness, "You will never be good." Mathematical research is important and it is fun. But so is having a family that loves you. I recall that I read a biography of a mathematician, perhaps it was Hilbert. The biography mentioned that when Hilbert heard that a graduate student had married, he would have nothing more to do with him. I think that was a common attitude back then that mathematics should take over your mind and your body. I bought into it and worked so hard.

## Graduate Research

Looking back at my research career I made a mistake. I never had a research program. I just loved working on problems. One of my best research tools was the question, "What are you working on?" I wrote several papers based on that question. Of course, I failed more often than I succeeded. I found out that I was not clever enough for this research style. That does not mean I was a failure. I earned a PhD in number theory and in the process participated in the mathematical adventures of a community, sometimes forging a trail myself, but mostly enjoying the scenic views provided by the rest of the community.

My advisor suggested that I work in the additive theory of groups and number theory. I studied his book on addition theorems and concluded that you had to work a great deal to make little advances. It did not appeal to me. Don Lawver, one of our faculty members, came to Mann and asked him if he knew the answer to a very simple question. Here is the set up. Let $F$ be a field of finite dimension over the field of rationals, $f(x)$ an irreducible polynomial over $F$, and $K = F(\theta)$, where $\theta$ is a root of $f(x)$. Let $O_F$ and $O_K$ be the rings of integers in the two fields and $P$ a prime ideal in $O_F$. A classical result in algebraic number theory shows that the prime ideal factorization $P$ in $O_K$ is obtained from factoring $f(x)$ modulo $P$, except for those $P$ that divide the discriminant of $f(x)$. Lawver's question was how does $P$ factor when $f(x) = x^p - a$, where $p$ is an odd prime and $P$ is a divisor of $p$ in $O_F$. In this case, $P$ divides the discriminant of $f(x)$ so the classical theorem does not apply. The case for $p = 2$ was well-known. Mann suggested that I investigate the question of how prime ideals factor in extensions of $F$ when a root of the irreducible polynomial $x^n - a$ is adjoined to $F$. This turned into my thesis problem.

While working on my thesis, I bumped into the irreducible binomial, $x^6 + 3$. When a root of this binomial is adjoined to $\mathbb{Q}$, the field of rationals, it yields a normal extension with Galois group $S_3$, the symmetric group. I mentioned this to Mann and this gave rise to a joint paper where we characterized all such normal binomials and their Galois groups. It turns out that Olga Taussky-Todd, a classmate of Mann in Vienna, had obtained this result many years before.

---

[8] a fifteenth birthday coming-out-into-society party

Mann said that I had to learn $p$-adic methods. Modern mathematics is written in this language. He did not use this language because his thesis advisor, Furtwangler, had a fight with Hensel, who developed $p$-adic methods. Mann, being Furtwangler's student, would not use $p$-adic methods. Later on Mann suggested that I look at his own doctoral thesis. In order to solve my thesis problem, I had to completely understand the results of his thesis. With a German dictionary at my side, I plowed through his thesis. He had left one case unsolved, and this case was critical for my work. Using $p$-adic completions, I resolved this last case, thus completing his thesis, and was able to move forward with my problem. I still remember the conversation that we had on this. I mentioned how I solved the problem by showing that a particular root of unity existed in the $p$-adic completion. He completely understood. The year after that, he included $p$-adic methods in his algebra course.

## Professional Career

In the summer of 1974, I was offered internships at Jet Propulsion Labs, Sandia Labs and Bell Labs. I selected Bell Labs in Murray Hill, NJ because Bernice and I had never been east of Tucson. Family and friends were concerned about this decision. If the children got sick, "were there doctors in New Jersey?" Others thought that Bernice should stay in Tucson with the children. "Was there food we could eat there?" "Where would we buy tortillas?" The question of our separation was never considered between us. It was an adventure. We bought an old tent-trailer and set off for a 3000-mile camping trip from Tucson to New Jersey with our three-year old, Ana, and six-week-old baby, Andrés. Storms caught us in Kansas. Having to get up in the middle of the night, turn on the camp stove to heat up Andrés's bottle was a nightly chore. When we returned home, we mentioned these incidents of hardship at a family gathering. One of the nanas of the family recounted her trip from Arizona to Sonora, Mexico when she was a 17-year old, traveling in a horse-drawn wagon with her sick child. That stopped our complaining. We had a wonderful time that summer as a family. Bell Labs was an amazing experience and I wrote a paper on a number-theoretic conjecture of Ron Graham.

When I completed my PhD, I did not want to be a professor. In fact, I did not apply for any jobs. Sandia Labs in Albuquerque, New Mexico, called me and asked if I would be interested in a position there. I visited and they hired me. I worked on the Command and Control of Atomic Weapons for two years. It was very interesting. In Albuquerque, the Society for the Advancement of Chicanos and Native Americans in Science (SACNAS) was just getting off the ground, and I joined many Chicano and Native American scientists in helping to form this organization. I later served as its president.

Many of us learn to deal with racism that occurs in our lives. We just push on. Institutionalized racism is much more difficult to deal with and academia is replete with it. It can be difficult to uncover this kind of racism, but it is there. In the late 1980s, I applied to work at the National Security Agency (NSA). In an exit interview, I was told that if I worked at NSA I could not have contact with foreign nationals. I replied that I worked at a university and this was impossible to comply with. The interviewer asked, "besides the university?" I am a Chicano, living 60 miles from the Mexican border, with relatives still living there. How could I avoid such contact? In the end, I said that I could not comply with this rule and I was turned down. I was a Vietnam veteran, had held a security clear-

Photo courtesy of W. Y. Velez.

Andrés, Ana, Bernice, Bill.

ance while I worked at Sandia Laboratories, yet this was not enough. It is perfectly fine for Chicanos to give their lives for this country, but it is not fine to devote our lives to work for this country.[9],[10]

After two years at Sandia, I found that I missed the teaching. I called Larry Grove, one of my professors at the University of Arizona (UA) to ask if there was a position available there. The department head called and asked me to give a lecture. I was hired and began as an assistant professor there in 1977. For the next ten years, I was lost in thought. There was a strong group in algebra and number theory. Dan Madden, who had been hired the year before me, had common mathematical interests and we investigated number-theoretic questions together. I worked so hard, though work is not a good description. It is not work when you love what you are doing. I didn't worry about tenure because I knew I would get it. However, like so many minority scientists, I felt out of place in the department, especially in the beginning.

Mathematically, I could not have asked for a better environment. But I found the social interactions with faculty draining. I have often said that just because we are serious it does not mean we should be somber. Where was the joy that we should feel in having the opportunity of being mathematicians and professors? People were very brusque with each other, and the etiquette of polite Mexican society was totally absent. At the beginning, Bernice and I hosted lots of social gatherings at our home. On one evening we had three mathematicians at dinner. Bernice would ask a question and they would answer me. They

---

[9] NSA Policy on Contact with Foreign Nationals, Letters to the Editor, *Notices of the American Math. Soc.*, Vol. 42 (2), February 1995, pg. 219.

[10] Names on a Wall—A Perspective on Why Diversity Matters, *American Scientist*, Vol. 85(2), March-April 1997, p. 200. 13.

couldn't talk to women. Such socializations were too much for both of us and we withdrew. Having one mathematician in the house was enough for Bernice.

The lack of socialization skills of mathematicians is legendary. If the mathematician looks at his own shoes when talking, he is in an introvert. When he looks at your shoes he is an extrovert. Is it their vision of how mathematicians should behave? Is this behavior genetic or learned? Do mathematicians feign such behavior to impress upon their students how deep in thought they are and how brilliant they are? I personally think this is a charade for most. Why can't our model for a brilliant mathematician be someone like Irving Kaplansky?[11] When I spoke with him he would bubble over with enthusiasm.

## On Being the First and Supporting Students

I was the first Chicano hired in a tenure-track position in the mathematics department at the University of Arizona (UA). Somehow this made me an expert on minority issues. I was trained to be a mathematical researcher yet expected to function as a sociologist. This is the common plight of minorities in our profession. When you are the only minority, the community views you differently. Perhaps you are viewed as being representative of your culture, like a zoo exhibit, or you are viewed as an expert witness for your culture and asked a puzzling number of questions. How do minorities prepare themselves for these encounters? This places an extra burden on us that the majority population does not have. It took me years to find my minority voice, and in this, SACNAS was a tremendous asset. It is a sign of the times that minority scientists have to come together to support each other. That is why organizations like SACNAS, National Association of Mathematicians, Association for Women in Mathematics came into existence. Shouldn't the mathematical community ask itself why women and minorities have to expend so much energy in creating these organizations to protect themselves from the majority?

In the 1980s, I became faculty advisor to the student chapter of the Society of Hispanic Professional Engineers. I attended most of their weekly meetings and their annual conferences. I began to learn how to advise students. This was the beginning of the end of my research career. I began more serious efforts at increasing minority participation in mathematics and after a few years it consumed me. I was not happy giving up my research career, but I was not bright enough to carry out this work with students and work on research.

I was very fortunate to come into contact with Phil Kutzko and the Math Alliance. I served on the board of the Math Alliance for many years. When I was president of SACNAS, I tried to institute a program in SACNAS to help minority students apply to graduate school. We tried it for a few years, but SACNAS was too big. The Math Alliance allowed me to propose this program again and the Facilitated Graduate Application Process program is now an important component of the Math Alliance.

Passion for life has always been part of my being. I passionately studied mathematics in college and taught with passion when I became a faculty member. When I decided that the mathematics major was the best major for students I passionately promoted the mathematics major, first among minority students. Then when given the opportunity to

[11] Irving Kaplansky (1917–2006) was a Canadian American mathematician, college professor, author, and musician.

direct the Math Center in the mathematics department, I directed this energy towards all students.

I am truly fortunate to have had a career in mathematics. I enjoyed all of it. I loved the research, the joy of teaching, and the challenges to convince students to take more mathematics. I lived an idyllic life, one that I shared with a beautiful person, Bernice.

# 27

# Dr. María Cristina Villalobos

In this chapter I recount my personal journey from K–12 schooling to my present position as faculty teaching at a university, I detail an important principle instilled in me by my mother—to take the initiative. Growing up, my mother would inform my two siblings and me to speak up and inquire about opportunities. By taking the initiative, we would be carving out a path to do better in our studies, which would lead to our future careers. In this narrative, I provide some of the many instances where I recall "taking the initiative." Currently, I still "take the initiative" in my own career, as I've learned that if I don't ask or if I don't try, then I end up doubting what possibilities could have occurred. "Taking the initiative" begins by having confidence in yourself and always inquiring about

Dr. María Cristina Villalobos

Illustration created by Ana Valle.

opportunities. And this is the principle that I now communicate to students and which centers my *testimonio*.

## My Upbringing

As is common in Mexican culture, I was named after my mother, while my younger brother was named after my father. My sister, who is the middle sibling, took on the name Gabriela, as it was a favorite name of my mother's. And thus my story begins. My parents, Jesús Villalobos Cuéllar and María Cristina Sánchez Treviño, were born and raised in Hualahuises, Nuevo León, a small town in northern Mexico, which is close to the larger town, Linares. My parents met at a dance hall in 1969 and married nine months later when my mother was 23 years old and my father was 44 years old.

My father was born in 1925 while my mother was born in 1946, and as you'll note they had an age difference of 21 years. Both of my parents came from large families consisting of eight to nine children. Life was difficult in Mexico as my grandfathers worked whatever jobs were available. My grandmothers were homemakers, taking care of the family at home. Hence, my father only completed elementary education through the third grade so that he could work and assist his family financially. He read the newspaper daily and could

Photo courtesy of Maria Cristina Villalobos.

With my family, sister and parents celebrating my father's 93rd birthday in November 2018.

engage in political conversations and debates with anyone. Now, imagine being nine years old and dropping out of school to take on an important matter that adults are tasked to take. Unfortunately, this still occurs in many parts of Latin America and around the world.

When he grew older, he crossed the Mexican-U.S. border in search of better jobs to support his family. I recall him telling me his adventures and the multiple times he was deported to Mexico. He remembered vividly the discrimination that he encountered in Victoria, Texas, at a barber shop where he could not get a haircut because he was Mexican. So he just walked away. Another incident occurred in McAllen. He was apprehended since he was without *papeles*.[1] He was put in the back of a police car and took the opportunity to flee when the two policemen decided to have lunch at a restaurant. Perhaps it was a hint for my father to leisurely walk away. Eventually, my father and his younger brother made their way to Chicago to work in a printing company and became U.S. residents. During this time, my parents stayed in contact by writing to each other. My uncle got married, had a family and settled in Chicago where he still lives. His children, my cousins, attended college, some completed graduate degrees, and all are successful in their careers.

My father returned to Hualahuises and married my mother in 1970. They immigrated to McAllen, Texas, in the Rio Grande Valley, a location close enough to their homes but most importantly a location they thought would provide a "better life." Soon after, my sister and I were born in McAllen and we all settled in Donna, a smaller town 12 miles

---

[1] The word *papeles* translates to "papers" and is used to refer to documentation establishing legal status in the U.S.

from McAllen. It was in Donna where I did all of my K–12 education. At that time there was only one elementary school, one middle school, and one high school; now there are multiples of each. During my school-aged years, my father worked at a local cannery in Donna while my mother worked at a local cannery in San Carlos and then as a custodian at the local Headstart in Donna.

**Early Schooling**. I was raised in a Spanish-speaking environment as Spanish was my parents' native language. My mother has always been a wise woman (as all Latina women are!) and a strong supporter of education. Thus, she took it upon herself to enroll me and my sister in the Donna Headstart preschool program. I spent one year in Headstart while my sister spent two given that she was a year younger than me. I only remember a few things about that year. I remember a field trip where I sprained my ankle, and singing while the teacher played the piano. I also remember watching a world slowly spin on the television screens, which years later I discovered was the intro to the soap opera *As the World Turns* that our teacher would watch while we were (supposed to be) napping.

From kindergarten through second grade, I was in bilingual education with bilingual teachers, perfecting Spanish and learning English. Developing into a dual-language speaker was helped by my immersion in Spanish at home and in English while in school. In third grade, I encountered my first full-blown English-speaking class—I say this since Ms. Turner only spoke English. I remember initially being frightened that Ms. Turner only spoke English, since I had felt comfortable with my bilingual teachers knowing that if needed to I could combine Spanish and English to communicate my thoughts. And thus I would frame English sentences with proper grammar in my mind before I approached Ms. Turner. Eventually this practice went away and the third grade was the best year of all my K–12 years. What made that year great was that we did so much "active learning," to put it in today's pedagogical use of words. In Ms. Turner's class we sang and danced to songs by the Beatles; we built a western town out of construction paper and straws and I was designated "sheriff"; and we built a vegetable garden whose produce we ate at an end-of-school year Hawaiian celebration, which was where I discovered that I didn't like radishes.

In terms of mathematics, this was the grade where we learned the multiplication tables. Ms. Turner would provide us with a new set of flashcards each time we mastered learning the multiples of a set of numbers. Most importantly, I learned that I was good at spelling. Ms. Turner would line up the class in the front of the classroom and provide each of us a word to spell orally. If we missed it once, we were out. I kept missing words and wanted to advance and also win first prize. One day I noticed Ms. Turner reading words from the back of our spelling book during our spelling orals. So I went home and studied the words and specifically remember being given the word "different" to spell to which I remembered

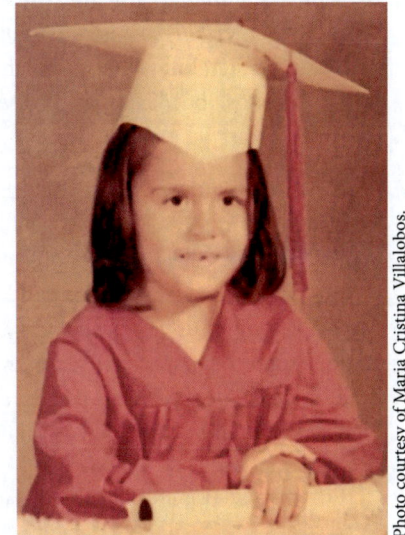

Kindergarten graduation.

Photo courtesy of Maria Cristina Villalobos.

not to forget the first "e." After that I won every oral spelling competition and brought home many prizes, one of which was a spider plant!

I grew up in a low-income family and mom always made it a point to tell us that she could buy us any food (minus candy) we wanted or educational materials, but not toys as they were reserved for birthdays and Christmas. So as a child I would request mathematics workbooks from our local grocery store, which allowed me to practice and learn mathematics.

**Middle school years**. Middle School turned out to be a fun time. In eighth grade, I studied algebra for the first time, but I found it to be difficult as the teacher didn't explain the concepts very well. During one weekend, I remember that my friends and I didn't understand a concept associated with a word problem. Some of my friends asked their parents for help since they had taken algebra in college; mind you this was a small minority. When my mother saw me struggling she immediately called her friend in Hualahuises who was a mathematics teacher. Guess what? We hopped on a bus that Saturday morning and made the four-hour ride to Hualahuises where Maestra Armandina devoted her afternoon to helping me learn algebra. Wow! Every time that I look back at this instance, I am in awe at my mother's courage and determination to help me succeed. Although she may not have taken algebra, she showed me that it is not impossible to learn and that there is always a way. The following day I explained the solution to my friends.

I had taken Spanish classes throughout middle school, learning the pronoun *vosotros*[2] and the verb tenses associated with it. I couldn't understand why we needed to learn that pronoun as it was neither used in the Valley nor in Mexico. During this time, I had become more interested in learning about where to place the accent mark in a Spanish word. I had asked my mother about the accent in the words *lápiz*[3] and *árbol*[4] to which she replied *solamente sabes*, that is, "you know." I didn't take that at face value as I figured there must be a systematic way, otherwise people would be placing accent marks left and right. When I took (another) Spanish class in eighth grade, my eyes lit up when the topic of accent marks came up. There were rules! I memorized those rules, borrowed my dad's newspaper *El Porvenir* and applied the rules. Yes, I had learned where accent marks belonged. In fact, I now knew why my first name "María" had an accent, but not "Cristina" nor "Villalobos," but "Sánchez" did. Accent marks had now been added to my world, and I was elated!

**High school years**. Having thought that I was weak in algebra, I decided to first take Geometry in ninth grade. I enjoyed the class as Mr. Frazier taught us to prove properties about angles, parallel lines, congruency, and other topics by using a two-column table format. Then came tenth grade and I enrolled in Algebra 2, which I feared. However, Mr. Miller taught us very well and I attended morning tutoring sessions to make sure that I understood the material well. Later I took Precalculus in eleventh grade, and in twelfth grade I took Calculus with Ms. Mendiola who turned out to be an exemplary teacher and who prepared me so well that I earned an "A" in first-semester Calculus I at University of

---

[2] The pronoun *vosotros* can be translated to "you all" in English.

[3] Pencil.

[4] Tree.

Texas at Austin. In fact, my strong high school preparation meant my first year in college was manageable and doable.

During high school, I learned to "Take the Initiative," which is what I now tell my students. In order to graduate high school with honors, one needed to take five honors classes, but you needed to pass a standardized test in order to enroll; note that AP courses or dual-enrollment courses were not available then. I was not good at taking standardized tests. I spoke to my mom about this issue, and she advised me to talk with the counselor. So that summer before eleventh grade when registering for classes, I took it upon myself to talk with the counselor. I mentioned that I could do the work of an honors class and I could obtain As in the classes and hence do better than some of the students who had tested in, but had gotten Cs. The counselor looked at my past grades and was surprised as to why I had not been in the honors courses. All said, I had to take five honors classes in two years to graduate with honors. And I did. I enrolled in the honors courses English III, English IV, Spanish 3 (yes there was a lot of Spanish to learn!), Government, Marine Biology, Anatomy and Physiology, and Calculus, and thus I graduated with honors in the top 1% of the class, which had approximately 350 students.

During high school I decided to participate in the University Interscholastic League (UIL) Spelling competition since I was good at spelling. We were provided with a booklet of over 1000 words to memorize and we competed locally in preparation for the District meet. I did very well. When I had trouble remembering the spelling of words, I would pronounce the word as if it were in Spanish as Spanish has only five vowel sounds, where-as English has five vowels, but with multiple sounds making it tricky to distinguish, for ex-ample, between a long "a" and a short "a" and everything else in between. So Spanish was a tool that I used to memorize the spelling of some challenging English words. I did very well placing in the top three in District every year and making it to Regionals to compete in San Antonio, but I never placed in the top three as it was tough.

In tenth grade we travelled with the UIL Typing team to San Antonio and I learned that no one had ever placed in the top three in a UIL State meet and the UIL Typing sponsor was eager for someone to do it; overhearing their conversation, I learned that the students were typing 45–55 words/minute with 0–2 errors. Wow! I had taken a typing course in middle school and I was typing 60 words/minute with errors. That is when I learned that students who competed at the UIL State meet were eligible to apply for a UIL college scholarship. So you can imagine what I did! I enrolled in a typing course in eleventh grade to be eligible to compete in UIL. I was earning first prizes through-out the local competitions until we got

Photo courtesy of Maria Cristina Villalobos.

Wearing my UIL State medal with my brother, May 1988.

Donna High School graduation, May 1989.

to the last practice where I dropped to second place. All of this time I hadn't quite paid attention to the advice from my UIL Typing sponsor, Ms. Medrano. She had advised me to seek a consistent typing pattern and hence I began to practice quickly. All told, I achieved second place at the UIL State meet at UT-Austin, becoming the first individual to place at the State level from Donna High School. Today these typing skills have certainly helped me greatly in writing grant proposals, research papers, letters of recommendations for students, and answering emails. These UIL trips to San Antonio and Austin were the first I had taken away from my family and away from the Valley. They opened my eyes to the rest of Central Texas.

During two high school summers, I enrolled in the Texas PreFreshmen Engineering Program (TexPREP) at our then regional university, Pan American University, which is now part of University of Texas Rio Grande Valley. I met many peers who became good friends and I took classes such as Logic, Computer Science, Engineering, and Physics, which helped me understand a bit more of what an engineering career entailed. In addition, one summer I decided to take on a part-time job where I worked as a shoe department clerk (for one month) and as a cashier in a grocery store. I took these jobs as I didn't know what awaited me in college especially since I would be a first-generation college student. I didn't actually know if I wanted to go to college. After these jobs and TexPREP, I figured I'd go to college to become a high school mathematics teacher.

## College Years

Knowing that I would graduate in the top 10% of my high school class, I knew that I had automatic admission to UT-Austin. I had already visited the campus during the UIL State meet and hence I decided to go there for college. In 1989, my freshmen year I took Calculus 1 and I learned that some peers were part of the Emerging Scholars Program (ESP). I wanted to get the best preparation since I was planning to become a high school math teacher and hence I "took the initiative" and inquired about becoming a part of ESP. The following semester I was part of ESP while taking Calculus 2. And I am glad I was in ESP. I encountered new and challenging material in Calculus 2. ESP met three times a week for two hours each time, providing a total of six hours of weekly meeting time where we collaboratively worked on problems. Outside of class, I met very often with my Graduate Teaching Assistant Ms. Jackie Bacon. The following year ESP offered me the opportunity to be a Student Teaching Assistant and help the current Calculus 2 students with the material. I was assigned to work with Graduate Teaching Assistant James Mendoza Álvarez who became a mentor during college and who is now a friend and colleague teaching at The University of Texas at Arlington in the Department of Mathematics.[5]

I will change direction here for a moment. UT and the city of Austin opened up my eyes to a different culture. No longer was I, a Latina, in the majority as I was in the Valley. Now I knew what it meant to be in the "minority." There were few Latinos in Austin, few Latino professors at UT, and few Latino students studying careers in STEM (science, technology, engineering, and mathematics). This was a shock to me. In addition, I met new Latino friends who spoke little Spanish, but it shouldn't have shocked me as they were multi-generational Latinos while I was first-generation Mexican-American. Moreover, I also learned that I had grown up in a low-income family as I compared myself to others. All of this time growing up in the Valley, I thought I was middle-class. Probably because my parents had bought a 25-year old home when I was in first grade and paid it off within four years (something my mother is extremely proud of!); because we had food on the table although we did ask our neighbor for a loan of $50 every now and then to make do with the week's expenses; because on a few occasions we purchased new furniture and clothes using the store's lay-away plan; and because compared with my cousins in Mexico we were doing financially better. There is no shame in being low-income, and I take great pride in the fact that my parents instilled in us the diligence of a strong work ethic and honesty along with "living within your means" to avoid debt, to pay back any loans in the event you needed to ask the same individual for a new loan, and to financially plan ahead. My siblings practice these ideals and are very successful in their own careers.

Now, in my second year at UT-Austin, I was advised to take a class with Dr. Efraim Armendáriz, who was chair of the mathematics department and a proponent of the ESP program. So I enrolled in his linear algebra class. I was so thrilled to see a Mexican-American mathematician and he was the only one in the department. In addition, he had spent some childhood years in Brownsville, Texas, a city in the Valley. Hence, we were connected! During my undergraduate years at UT-Austin, a school of over 50,000 college students, Dr. Armendáriz and Dr. Uri Treisman became my mentors. On the recommendation of Dr. Armendáriz, a cohort of Mexican-American and African-American students applied to a summer research program at University of California, Berkeley which was led by Dr. Treisman. After that summer, Dr. Treisman joined the faculty at UT-Austin and I took his class on modern algebra, where I learned how to write proofs, and a second course on Galois theory. Both mentors encouraged me to collaborate with two students on a math project and we presented our work at the 1992 Society for the Advancement of Chicanos/Hispanics and Native Americans in Science (SACNAS) national conference held in Chicago. There I met Dr. Bill Velez,[6] a mathematician at the University of Arizona, who has been a mentor, colleague, and friend. He had started a conversation with me where I shared that I was planning to walk to The Art Institute of Chicago as I had taken an art history class in college and I was eager to view paintings and sculptures that I had studied and apply my knowledge of distinguishing Byzantine and Renaissance paintings and everything in between. He joined me and we spent an afternoon together at the museum.

While at UT-Austin, a friend had shared a newspaper clipping of another Mexican-American mathematician—Dr. Richard Tapia[7] from Rice University who had just been

---

[5] Dr. James A. Mendoza Álvarez is featured in Chapter 1.

[6] Dr. William Yslas Velez is featured in Chapter 26.

[7] Dr. Richard Tapia is featured in Chapter 22.

UT-Austin graduation with my mentor Dr. Efraim Armendáriz, chair of the mathematics department, May 1994.

inducted into the National Academy of Engineering. What an honor! My friend contacted Dr. Tapia who invited us to visit him. And thus another adventure began. We "took the initiative" and drove to Houston one Saturday morning in March/April 1993 to visit him. We had dinner at Picos Restaurant with Dr. Tapia and his family, Dr. Virginia Torczon who is now faculty at William and Mary, and Dr. Michael Trosset, who is now faculty at the Indiana University at Bloomington and with whom I have become good friends. During dinner we discussed mathematics and chatted about life. When I returned to Austin, I called Dr. Tapia and asked if I could apply to his Spend a Summer with a Scientist (SAS) research program although the application deadline had passed. He asked me to contact his assistant, Ms. Theresa Chatman, and submit an application. I was accepted and that summer I worked on a mathematics education project with Dr. Anne Papakonstantinou in the Rice University Summer Mathematics Program (RUSMP) alongside high school teachers. During SAS we had Friday meetings where participants presented their research work. I learned of optimization and differential equation applications through the work of doctoral graduate students, Tony Kearsley now at the National Institute of Science and Technology, Cassandra McZeal now at Exxon-Mobile, and Mónica Martínez now at Intel, and I found it amazing that I was learning of mathematics applications.

The following academic year, 1993–1994, was my senior year at UT-Austin and there was a lot I had to do. In fall 1993, I applied to applied mathematics graduate programs across the nation and I also applied for fellowships. It was a demanding semester, but I knew I needed to give my all. Since I had only one shot at the applications and fellowships, I did not seek part-time work that semester and instead took out a loan. In addition, I also attended the SACNAS conference where Mónica introduced me to Dr. Juan Meza from Sandia National Laboratory and by the end of the conference I had secured a summer internship at Sandia! My efforts that fall semester paid off, and I got accepted to the Computational and Applied Mathematics (CAAM) department at Rice University and I received a three-year Ford Foundation Predoctoral Fellowship. I was elated! And thus I spent the next five years at Rice University and in 2000 I received my PhD in CAAM.

To summarize, I had met many Latino mathematicians during my college years who mentored me, who continue to assist me in my career and who have become friends and family. Much of these efforts were attributed to "taking the initiative" and meeting them and learning about opportunities.

# Graduate School Years

My first year at Rice was smooth and I attribute it to the preparation that UT gave me. I was used to studying late into the night, working on proofs and perfecting them, and just plain working hard. I studied with friends and we quizzed ourselves on the material. I was doing well until I had a moment of truth. I was probably in my second year of graduate studies when I walked into Dr. Tapia's office telling him that I wasn't sure how I could be working behind a computer doing applied mathematics while people in other parts of the world were struggling to live day by day. So his advice was to integrate community service into one's career to start making positive changes. And thus I have done exactly that in broadening STEM participation to women and underrepresented minorities. At Rice, I was a tutor at a local middle school and I served as a mentor to other students. I was nominated by Dr. Tapia for the Rice Volunteer award, which I received.

If you were a part of Dr. Tapia's research group, you quickly learned about the "Torture Chamber," the place where you presented your research work only to be critiqued and questioned by your peers regarding the material. It actually sounds worse than it really was. My peers and I presented our work many times, more than I remember. But think about it like this: questioning is good, as it helps you develop into a researcher and it provides you with opportunities to cement your work, understand it better, and develop new directions of research. In my third year of graduate studies, Dr. Yin Zhang became my PhD advisor (with Dr. Tapia as a coadvisor) for my research work in optimization. Throughout graduate school, Dr. Tapia became my mentor, and throughout my professional career, he has become a colleague, friend, and family.

Apart from studying, graduate school was a lot of fun. There were Thanksgiving dinners and cookouts at the Tapia's home. And then there were the times when we made tamales at Dr. Tapia's home. We purchased masa, corn, chicken and other ingredients. I'll never forget when we made a phone call to Dr. Tapia's mother in Los Angeles to find out how we'd know if the masa was ready; she told us to place a piece of it in a glass of water and if it floated it was ready. So we had two winters when we had a *tamalada*[8] and the entire CAAM department was invited to Dr. Tapia's home to enjoy our tamales. Now the department had many Latino graduate students from Venezuela and Colombia and so it was critical that we learn how to dance salsa and merengue. Thus one summer about 15 students—all from diverse backgrounds—learned to dance salsa and merengue in Dr. Tapia's garage from the daughter of one of the graduate students from Colombia. Since then I have applied these dancing skills and they've become useful at weddings, social gatherings, and even at conference events!

There were few Latino graduate students across Rice and as a result we got to know each other and became friends. The CAAM department was very diverse in terms of Latino, Black, and female representation. Friday afternoon meetings with Dr. Tapia's group meant discussing not only research, but also social/educational justice issues, too. And it also meant having a better understanding of my friends who were Black, Brown, White, and all colors in between. During those meetings we shared our upbringing experiences and respected each others' opinions. Sure, some discussions were tough and emotional, but we learned from each other.

---

[8] A party where the main meal is tamales.

Photo courtesy of Maria Cristina Villalobos.

My friends and I at Disney World, July 1999. Out of the eight individuals pictured, five defended their doctoral dissertations in CAAM (two), Chemical Engineering, Economics, and Mechanical Engineering.

In summer 1999, I defended my dissertation successfully. My CAAM peers and friends attended and provided much support. Actually, several of my friends and I defended our doctoral dissertation within days of each other. So what do you do when you accomplish something great? You go to Disney World!

## Professional Career

As a tenured professor at The University of Texas Rio Grande Valley (UTRGV), I teach mathematics classes. I work with colleagues in electrical engineering, computer science, and applied mathematics to model application problems using optimization and optimal

Photos courtesy of Maria Cristina Villalobos.

(L) My then fiancé, now husband Arturo and I at our Rice PhD graduation in May 2000. Arturo also teaches at UTRGV in mechanical engineering. (R) Rice PhD graduation with my parents and Dr. Richard Tapia.

(L) My first faculty appointment at St. Mary's University, August 1999. (R) Currently, Full Professor at The University of Texas Rio Grande Valley, 2020, where I hold an endowed position.

control. I also collaborate with colleagues on improving STEM education, and providing service to the department, university, local community, and to the mathematics profession. I am the founding director of the Center of Excellence in STEM Education, which focuses on broadening STEM participation of women and underrepresented minorities, especially that of Latinos, and preparing them for their careers or graduate studies. Our Center was one of three funded nationwide and I am extremely proud of that accomplishment. In addition, I collaborate with colleagues on grant proposals, research publications, teaching initiatives to improve student success, and service activities. I enjoy my job. And I am passionate about preparing the next generation of students to become leaders in academia, government, and industry.

In 2015, I was appointed Interim Director of the School of Mathematical and Statistical Sciences to transition two mathematics departments into one department consisting of 45 tenure-track/tenured faculty and 25 lecturers to The University of Texas Rio Grande Valley. I was the first Latina and the first woman to serve as department chair in the history of both departments. During my two-year period as chair, I hired a total of nine tenure-track faculty, effectively increasing the number of Latino faculty by 66% and the number of women faculty by 40%. Given these large increases it is important to ask the base value; that is, the original numbers of Latinos and women on the faculty. I can tell you that it was merely a handful or less for each group.

Due to my leadership in STEM, I have received many national awards. The one that summarizes my accomplishments in mentoring students and faculty is the Presidential Award for Excellence in Science, Mathematics, and Engineering Mentoring (PAESMEM) which is one of the top honors awarded by the White House. The virtual ceremony took place in August 2020. I shared the excitement and award with my family and mother who were present. Once the pandemic[9] subsides, a formal ceremony will take place at the White House.

---

[9] This was written during the COVID-19 pandemic.

## Concluding Advice

As you've noticed I've had mentors during my school and college years, however these mentors have also extended to my professional career providing guidance and advise. And so I end by asking you to "Take the Initiative!" and knock on doors. Take things a step at a time, but look five steps ahead. Be proud of your heritage and hold your head high. And finally I couldn't leave without stating: *Do well in Mathematics!*

# About the Editors

**Dr. Pamela E. Harris** is a Mexican-American mathematician and Associate Professor in the Department of Mathematics and Statistics at Williams College. She received her AA and AS from Milwaukee Area Technical College, BS from Marquette University, and MS and PhD in mathematics from the University of Wisconsin-Milwaukee. Dr. Pamela E. Harris's research is in algebraic combinatorics and she is the author of over 50 peer-reviewed research articles in internationally recognized journals. An award-winning mathematical educator, Dr. Harris received the 2020 MAA Northeast Section Award for Distinguished College or University Teaching, the 2019 MAA Henry L. Alder Award for Distinguished Teaching

Dr. Pamela E. Harris

by a Beginning College or University Mathematics Faculty Member, the 2019 Council on Undergraduate Research Mathematics and Computer Sciences Division Early Career Faculty Mentor Award, was named a 2020 Inaugural Class of Karen EDGE Fellow, and was one of 50 women featured in the book *Power in Numbers: The Rebel Women of Mathematics*. Her professional mission is to develop learning communities that reinforce students' self-identity as scientists, in particular for women and underrepresented minorities. In support of this mission, Dr. Harris co-organizes research symposia and professional development sessions for the national conference of the Society for the Advancement of Chicanos/Hispanics and Native Americans in Science (SACNAS), and is an editor of the e-Mentoring Network blog of the American Mathematical Society. Moreover, in order to provide visibility to and increase the positive impact of the role models within our community, Dr. Harris co-founded Lathisms.org, a platform that features the contributions of Latinx and Hispanic scholars in the mathematical sciences. She cohosts the podcast "Mathematically Uncensored," sponsored by The Center for Minorities in the Mathematical Sciences, and has recently coauthored the book *Asked And Answered: Dialogues On Advocating For Students of Color in Mathematics*.

Dr. Alicia Prieto Langarica

**Dr. Alicia Prieto-Langarica** is a Distinguished Professor in the Department of Mathematics and Statistics at Youngstown State University. She obtained the Distinguished Professor designation in May 2020, after being awarded the distinguished professor award in Research, Teaching and Service. She received her Undergraduate degree in applied mathematics from the University of Texas at Dallas in 2008 and her PhD from the University of Texas at Arlington in 2012. Prieto-Langarica's research is in the intersection of mathematics and biology, specifically problems related to the medical field. Recently she started conducting research in data science, public policy and mathematics education. Some of her awards include the MAA Henry L. Alder Award for Distinguished Teaching by a Beginning College or University Mathematics Faculty Member, the 2020 SmithMurphy Award by the Student Government Association at YSU, the Athena Award Finalist by the Mahoning Valley Regional Chamber of Commerce, and The 25 Under 35 Mahonning Valley Young Professionals MVP award.

Dr. Vanessa Rivera Quiñones

**Dr. Vanessa Rivera Quiñones** is a mathematical biologist with a passion for telling stories through numbers using mathematical models, data science, and education. Born in Puerto Rico, her love for mathematics began at an early age and continued to grow thanks to the encouragement of her family, teachers, and the support of many mentors. She received her bachelor's (BS) degree from the University of Puerto Rico at Río Piedras (2013) and her doctoral degree (2019) from the University of Illinois at Urbana-Champaign in mathematics. She has been involved in several organizations and initiatives that focus on broadening the participation and mentoring of underrepresented students in mathematics. In 2015, she was awarded the Ford Foundation Predoctoral Fellowship, which seeks to diversify colleges and universities. She is a proud member of multiple national organizations such as the Association for Women in Mathematics (AWM), SACNAS, American Mathematical Society (AMS), Mathematical Association of America (MAA), and the National Alliance for Doctoral Studies in the Mathematical Sciences. She believes mathematics is a human endeavor and that by creating inclusive and equitable environments that embrace the identities of who does mathematics, our community will flourish. Currently, she is a data science consultant and instructor. At the next step of her career, she is interested in working on the ever-growing challenges of sustainability, healthcare, and education through a social justice lens.

**Dr. Luis Sordo Vieira** is a Research Assistant
Professor in the Department of Medicine at
The University of Florida. He is a Venezuelan-
American Applied Mathematician with scientific
experience broadly described as being in the area
of systems medicine. He completed his Bachelor
of Science from Wayne State University in mathematics, minoring in physics, and his PhD in
number theory from the University of Kentucky as
a National Science Foundation Graduate Student
Fellow. He has served in the Lathisms leadership
team since 2019. Dr. Sordo Vieira has served in

Dr. Luis Sordo Vieira

the organizing committee for The Mathematics Summer Workshop for Achieving Greater
Graduate Educational Readiness, a program to prepare students from underrepresented
minorities in mathematics for graduate studies. He is a Society for Industrial and Applied
Mathematics Science Policy Fellow and received the American Mathematical Society
Simons Foundation Travel Grant.

**Dr. Rosaura Uscanga Lomelí** was born in
Mexico and came to the U.S. at the age of 11, so
she considers herself a Mexican-American. She
is an Assistant Professor in the Department of
Mathematics and Computer Sciences at Mercy
College (as of Fall 2021). Her research area lies in
mathematics education, specifically in the teaching
and learning of abstract algebra. She completed her
PhD at Oklahoma State University in 2021. Her
dissertation explored students' thinking regarding
the concept of "function" in the context of abstract
algebra. She received her BS in mathematics from
The University of Texas at Arlington in 2012 and

Dr. Rosaura Uscanga Lomelí

her MS in mathematics from Oklahoma State University in 2015. Dr. Uscanga is passionate about teaching and enjoys working with students—one of the reasons she decided
to study mathematics education. She strives to make sure students in her classroom feel
a sense of belonging and view themselves positively in relation to mathematics. She is
extremely interested in issues of equity, diversity, and inclusion in mathematics.

**Dr. Andrés R. Vindas Meléndez** is a Costa Rican-American mathematician, raised
in Lynwood, South East Los Angeles, California. He is a first-generation college graduate
and is currently a National Science Foundation Postdoctoral Fellow at the University of
California, Berkeley and Mathematical Sciences Research Institute Postdoctoral Fellow.
He completed his PhD at the University of Kentucky where he was supported by a
National Science Foundation Graduate Research Fellowship and by a National Science

Dr. Andrés R. Vindas Meléndez

Foundation Louis Stokes Alliance for Minority Participation Bridge to Doctorate Fellowship. At the University of Kentucky he was also an affiliated graduate student in the Latin American Studies program and earned a graduate certificate in Latin American, Caribbean, and Latino/a Studies. He earned a master's degree in mathematics at San Francisco State University and completed his undergraduate degree in mathematics at the University of California, Berkeley where he also minored in Philosophy and Chicana/o & Latina/o Studies. His research interests are in algebraic, enumerative, and geometric combinatorics. In particular, he is interested in lattice-point enumeration for polyhedra. Dr. Vindas Meléndez's teaching, service, and outreach is student-centered. He has the opportunity to help guide students to learn abstract mathematics and find their voice while also developing a sense of ownership of their knowledge and mathematical abilities. Dr. Vindas Meléndez strives to create community in order to build students' confidence in spite of society's negative messages and stigma about mathematics. He also aims to build meaningful and empowering experiences with mathematics, while also challenging others to think about the power structures that are present in and outside mathematical spaces.